Innovative Quantum Computing

Online at: https://doi.org/10.1088/978-0-7503-5281-9

IOP Series in Coherent Sources, Quantum Fundamentals, and Applications

About the Editor

F J Duarte is a laser physicist based in Western New York, USA. His career has covered three continents while contributing within the academic, industrial, and defense sectors. Duarte is editor/author of 15 laser optics books and sole author of three books: *Tunable Laser Optics, Quantum Optics for Engineers*, and *Fundamentals of Quantum Entanglement*. Duarte has made original contributions in the fields of coherent imaging, directed energy, high-power tunable lasers, laser metrology, liquid and solid-state organic gain media, narrow-linewidth tunable laser oscillators, organic semiconductor coherent emission, *N*-slit quantum interferometry, polarization rotation, quantum entanglement, and space-to-space secure interferometric communications. He is also the author of the generalized multiple-prism grating dispersion theory and pioneered the use of Dirac's quantum notation in *N*-slit interferometry and classical optics. His contributions have found applications in numerous fields, including astronomical instrumentation, dispersive optics, femtosecond laser microscopy, geodesics, gravitational lensing, heat transfer, laser isotope separation, laser medicine, laser pulse compression, laser spectroscopy, mathematical transforms, nonlinear optics, polarization optics, and tunable diode-laser design. Duarte was elected Fellow of the Australian Institute of Physics in 1987 and Fellow of the Optical Society of America in 1993. He has received various recognitions, including the *Paul F Foreman Engineering Excellence Award* and the *David Richardson Medal* from the Optical Society.

Coherent Sources, Quantum Fundamentals, and Applications

Since its discovery the laser has found innumerable applications from astronomy to zoology. Subsequently, we have also become familiar with additional sources of coherent radiation such as the free electron laser, optical parametric oscillators, and coherent interferometric emitters. The aim of this book Series in Coherent Sources, Quantum Fundamentals, and Applications is to explore and explain the physics and technology of widely applied sources of coherent radiation and to match them with utilitarian and cutting-edge scientific applications. Coherent sources of interest are those that offer advantages in particular emission characteristics areas such as broad tunability, high spectral coherence, high energy, or high power. An additional area of inclusion are the coherent sources capable of high performance in the miniaturized realm. Understanding of quantum fundamentals can lead to new and better coherent sources and unimagined scientific and technological applications. Application areas of interest include the industrial, commercial, and medical sectors. Also, particular attention is given to scientific applications with a bright future such as coherent spectroscopy, astronomy, biophotonics, space communications, space interferometry, quantum entanglement, and quantum interference.

Publishing benefits
Authors are encouraged to take advantage of the features made possible by electronic publication to enhance the reader experience through the use of color, animation and video, and incorporating supplementary files in their work.

Do you have an idea of a book that you'd like to explore?
For further information and details of submitting book proposals, see iopscience.org/books or contact Ashley Gasque at ashley.gasque@iop.org.

A full list of titles published in this series can be found here: https://iopscience.iop.org/bookListInfo/series-in-coherent-sources-and-applications.

Innovative Quantum Computing

Steven Duplij
Center for Information Processing WWU IT, University of Münster,
48149 Münster, Germany

Raimund Vogl
Center for Information Processing WWU IT, University of Münster,
48149 Münster, Germany

IOP Publishing, Bristol, UK

ISBN 978-0-7503-5281-9 (ebook)
ISBN 978-0-7503-5279-6 (print)
ISBN 978-0-7503-5282-6 (myPrint)
ISBN 978-0-7503-5280-2 (mobi)

DOI 10.1088/978-0-7503-5281-9

Version: 20231101

IOP ebooks

British Library Cataloguing-in-Publication Data: A catalogue record for this book is available from the British Library.

Published by IOP Publishing, wholly owned by The Institute of Physics, London

IOP Publishing, No.2 The Distillery, Glassfields, Avon Street, Bristol, BS2 0GR, UK

US Office: IOP Publishing, Inc., 190 North Independence Mall West, Suite 601, Philadelphia, PA 19106, USA

Contents

Preface

This book is devoted to the study of exotic and non-standard mathematical methods in quantum computing. The principal ingredients of quantum computation are qubits and their transformations, which can be provided in different ways: first mathematically, and they can then be further realized in hardware.

In this book we consider various extensions of the qubit concept per se, starting from the obscure qubits introduced by the authors, and other fundamental generalizations. We then introduce a new kind of gate, higher braiding gates, which are implemented for topological quantum computations, as well as unconventional computing, when computational complexity is affected by its environment, which needs an additional stage of computation. Other generalizations are also considered and explained in a widely accessible and easy to understand style.

This book will be useful for graduate students and last year students for additional advanced chapters of lecture courses in quantum computer science and information theory.

Steven Duplij and Raimund Vogl
Münster, Germany
August 2023

Author biographies

Steven Duplij

Steven Duplij (Stepan Douplii) is a theoretical and mathematical physicist from the University of Münster, Germany. He was born in Chernyshevsk-Zabaikalsky, Russia, and studied at Kharkov University, Ukraine where he gained his PhD in 1983. While working at Kharkov, he received the title Doctor of Physical and Mathematical Sciences by Habilitation in 1999. Dr Duplij is the editor-compiler of the 'Concise Encyclopedia of Supersymmetry' (2005, Springer), and is the author of more than a hundred scientific publications and several books. He is listed in the World Directory of Mathematicians, Marques Who Is Who In America, the Encyclopedia of Modern Ukraine, the Academic Genealogy of Theoretical Physicists, and the Mathematics Genealogy Project. His scientific interests include supersymmetry and quantum groups, advanced algebraic structures, gravity and nonlinear electrodynamics, constrained systems, and quantum computing.

Raimund Vogl

Raimund Vogl is the CIO of the University of Münster (Germany) and has been the director of the University's IT center since 2007. He holds a PhD in elementary particle physics from the University of Innsbruck (Austria). After completing his PhD studies in 1995, he joined Innsbruck University Hospital as an IT manager for medical image data solutions and moved on to become the deputy head of IT. He is board member and president of EUNIS (European University Information Systems Organisation), and a member of Deutsche Gesellschaft für Medizinische Informatik, Biometrie und Epidemiologie (GMDS) and the Association for Information Systems (AIS). His current research interest in the field of information systems and information management focuses on the management of complex information infrastructures.

IOP Publishing

Innovative Quantum Computing

Steven Duplij and Raimund Vogl

Chapter 1

Obscure qubits and membership amplitudes

Nowadays, the development of quantum computing technique is governed by theoretical extensions of its ground concepts (Nielsen and Chuang 2000, Kaye *et al* 2007, Williams and Clearwater 1998). One of these extensions is to allow two kinds of uncertainty, sometimes called randomness and vagueness/fuzziness (for a review, see, Goodman and Nguyen 2002), which leads to the formulation of combined probability and possibility theories (Dubois *et al* 2000) (see, also, Bělohlávek 2002, Dubois and Prade 2000, Smith 2008, Zimmermann 2011). Various interconnections between vagueness and quantum probability calculus were considered in Pykacz (2015), Dvurečenskij and Chovanec (1988), Bartková *et al* (2017), and Granik (1994), including the treatment of inaccuracy in measurements (Gudder 1988, 2005), non-sharp amplitude densities (Gudder 1989), and the related concept of partial Hilbert spaces (Gudder 1986).

Relations between truth values and probabilities were also given in Bolotin (2018). The hardware realization of computations with vagueness was considered in Hirota and Ozawa (1989), and Virant (2000). On the fundamental physics side, it was shown that the discretization of space-time at small distances can lead to a discrete (or fuzzy) character for the quantum states themselves.

With a view to applications of these ideas in quantum computing, we introduce a definition of quantum state that is described by both a quantum probability and a membership function (Duplij and Vogl 2021), and thereby incorporate vagueness/fuzziness directly into the formalism. In addition to the probability amplitude, we will define a membership amplitude, and such a state will be called an obscure/fuzzy qubit (or qudit) (Duplij and Vogl 2021).

In general, the Born rule will apply to the quantum probability alone, while the membership function can be taken to be an arbitrary function of all of the amplitudes fixed by the chosen model of vagueness. Two different models of obscure-quantum computations with truth are proposed below: (1) a Product obscure qubit, in which the resulting amplitude is the product (in $\mathbb{C}$) of the quantum

doi:10.1088/978-0-7503-5281-9ch1

amplitude and the membership amplitude; and (2) a Kronecker obscure qubit, for which computations are performed in parallel, so that quantum amplitudes and the membership amplitudes form vectors, which we will call obscure-quantum amplitudes. In the latter case, which we call a double obscure-quantum computation, the protocol of measurement depends on both the quantum and obscure amplitudes. In this case, the density matrix need not be idempotent. We define a new kind of gate, namely, obscure-quantum gates, which are linear transformations in the direct product (not in the tensor product) of spaces: a quantum Hilbert space and a so-called membership space having special fuzzy properties (Duplij and Vogl 2021). We then introduce a new concept of double (obscure-quantum) entanglement, in which vector and scalar concurrences are defined and computed for concrete examples.

1.1 Preliminaries

To establish a notation standard in the literature (see, e.g. Nielsen and Chuang 2000, Kaye *et al* 2007), we present the following definitions. In an underlying d-dimensional Hilbert space, the standard qudit (using the computational basis and Dirac notation) $\mathcal{H}_{\rm q}^{(d)}$ is given by

$$| \psi^{(d)} \rangle = \sum_{i=0}^{d-1} a_i \mid i \rangle, \quad a_i \in \mathbb{C}, \mid i \rangle \in \mathcal{H}_{\rm q}^{(d)}, \tag{1.1}$$

where a_i is a probability amplitude of the state $| \, i \rangle$. (For a review, see, e.g. Genovese and Traina 2008, Wang *et al* 2020.) The probability $_{pi}$ to measure the ith state is $p_i = F_{p_i}(a_1, \dots, a_n)$, $0 \leqslant p_i \leqslant 1$, $0 \leqslant i \leqslant d-1$. The shape of the functions F_{p_i} is governed by the Born rule $F_{p_i}(a_1, \dots, a_d) = | \, a_i \, |^2$, and $\sum_{i=0}^{d} {}_{p_i} = 1$. A one-qudit ($L = 1$) quantum gate is a unitary transformation $U^{(d)}\colon \mathcal{H}_{\rm q}^{(d)} \to \mathcal{H}_{\rm q}^{(d)}$ described by unitary $d \times d$ complex matrices acting on the vector (1.1), and for a register containing L qudits quantum gates are unitary $d^L \times d^L$ matrices. The quantum circuit model (Deutsch 1985, Barenco *et al* 1995) forms the basis for the standard concept of quantum computing. Here the quantum algorithms are compiled as a sequence of elementary gates acting on a register containing L qubits (or qudits), followed by a measurement to yield the result (Lloyd 1995, Brylinski and Brylinski 1994).

For further details on qudits and their transformations, see for example the reviews by Genovese and Traina (2008) and Wang *et al* (2020) and the references therein.

1.2 Membership amplitudes

Innovation 1.1. *We define an obscure qudit with d states* via *the following superposition (in place of that given in (1.1))*

$$\left| \psi_{\rm ob}^{(d)} \right\rangle = \sum_{i=0}^{d-1} \alpha_i a_i \mid i \rangle, \tag{1.2}$$

where a_i is a (complex) probability amplitude $a_i \in \mathbb{C}$, and we have introduced a (real) membership amplitude α_i, with $\alpha_i \in [0, 1]$, $0 \leqslant i \leqslant d - 1$.

The probability p_i to find the ith state upon measurement and the membership function μ_i (of truth) for the ith state are both functions of the corresponding amplitudes, as follows

$$p_i = F_{p_i}(a_0, \dots, a_{d-1}), \quad 0 \leqslant p_i \leqslant 1, \tag{1.3}$$

$$\mu_i = F_{\mu_i}(\alpha_0, \dots, \alpha_{d-1}), \quad 0 \leqslant \mu_i \leqslant 1. \tag{1.4}$$

The dependence of the probabilities of the ith states upon the amplitudes, i.e., the form of the function F_{p_i} is fixed by the Born rule

$$F_{p_i}(a_1, \dots, a_n) = | a_i |^2, \tag{1.5}$$

while the form of F_{μ_i} will vary according to different obscurity assumptions. In this paper we consider only real membership amplitudes and membership functions—complex obscure sets and numbers were considered in Buckley (1989), Ramot *et al* (2002), and Garrido (2012). In this context, the real functions F_{p_i} and F_{μ_i}, $0 \leqslant i \leqslant d - 1$ will contain complete information about the obscure qudit (1.2).

We impose the normalization conditions

$$\sum_{i=0}^{d-1} p_i = 1, \tag{1.6}$$

$$\sum_{i=0}^{d-1} \mu_i = 1, \tag{1.7}$$

where the first condition is standard in quantum mechanics, while the second condition is taken to hold by analogy. Although (1.7) may not be satisfied, we will not consider that case.

For $d = 2$, we obtain for the obscure qubit the general form, instead of that in (1.2),

$$\left| \psi_{\text{ob}}^{(2)} \right\rangle = \alpha_0 a_0 \,|\, 0\rangle + \alpha_1 a_1 \,|\, 1\rangle, \tag{1.8}$$

$$F_{p_0}(a_0, a_1) + F_{p_1}(a_0, a_1) = 1, \tag{1.9}$$

$$F_{\mu_0}(\alpha_0, \alpha_1) + F_{\mu_1}(\alpha_0, \alpha_1) = 1. \tag{1.10}$$

The Born probabilities to observe the states $| 0\rangle$ and $| 1\rangle$ are

$$p_0 = F_{p_0}^{\text{Born}}(a_0, a_1) = | a_0 |^2, \quad p_1 = F_{p_1}^{\text{Born}}(a_0, a_1) = | a_1 |^2. \tag{1.11}$$

Innovation 1.2. *The membership functions are*

$$\mu_0 = F_{\mu_0}(\alpha_0, \alpha_1), \quad \mu_1 = F_{\mu_1}(\alpha_0, \alpha_1). \tag{1.12}$$

If we assume the Born rule (1.11) for the membership functions as well

$$F_{\mu_0}(\alpha_0, \alpha_1) = \alpha_0^2, \quad F_{\mu_1}(\alpha_0, \alpha_1) = \alpha_1^2, \tag{1.13}$$

which is one of various possibilities depending on the chosen model, then

$$|\, a_0 \,|^2 + |\, a_1 \,|^2 = 1, \tag{1.14}$$

$$\alpha_0^2 + \alpha_1^2 = 1. \tag{1.15}$$

Using (1.14)–(1.15) we can parameterize (1.8) as

$$\left|\, \psi_{\mathrm{ob}}^{(2)} \right\rangle = \cos\frac{\theta}{2}\cos\frac{\theta_\mu}{2}\,|\, 0\rangle + e^{i\varphi}\sin\frac{\theta}{2}\sin\frac{\theta_\mu}{2}\,|\, 1\rangle, \tag{1.16}$$

$$0 \leqslant \theta \leqslant \pi, \quad 0 \leqslant \varphi \leqslant 2\pi, \quad 0 \leqslant \theta_\mu \leqslant \pi. \tag{1.17}$$

Therefore, obscure qubits (with Born-like rule for the membership functions) are geometrically described by a pair of vectors, each inside a Bloch ball (and not as vectors on the boundary spheres, because '$|\sin|, |\cos| \leqslant 1$'), where one is for the probability amplitude (an ellipsoid inside the Bloch ball with $\theta_\mu = \mathrm{const}_1$) and the other is for the membership amplitude (which is reduced to an ellipse, being a slice inside the Bloch ball with $\theta = \mathrm{const}_2$, $\varphi = \mathrm{const}_3$). However, the norm of the obscure qubits is not constant because

$$\left\langle \psi_{\mathrm{ob}}^{(2)} \middle| \psi_{\mathrm{ob}}^{(2)} \right\rangle = \frac{1}{2} + \frac{1}{4}\cos\left(\theta + \theta_\mu\right) + \frac{1}{4}\cos\left(\theta - \theta_\mu\right). \tag{1.18}$$

In the case where $\theta = \theta_\mu$, the norm (1.18) becomes $1 - \frac{1}{2}\sin^2\theta$, reaching its minimum $\frac{1}{2}$ when $\theta = \theta_\mu = \frac{\pi}{2}$.

Note that for complicated functions $F_{\mu_{0,1}}(\alpha_0, \alpha_1)$, the condition (1.15) may be not satisfied but the condition (1.7) should nevertheless always be valid. The concrete form of the functions $F_{\mu_{0,1}}(\alpha_0, \alpha_1)$ depends upon the chosen model. In the simplest case, we can identify two arcs on the Bloch ellipse for α_0, α_1 with the membership functions and obtain

$$F_{\mu_0}(\alpha_0, \alpha_1) = \frac{2}{\pi}\arctan\frac{\alpha_1}{\alpha_0}, \tag{1.19}$$

$$F_{\mu_1}(\alpha_0, \alpha_1) = \frac{2}{\pi}\arctan\frac{\alpha_0}{\alpha_1}, \tag{1.20}$$

such that $\mu_0 + \mu_1 = 1$, as in (1.7).

In Mannucci (2006) and Maron *et al* (2013) a two stage special construction of quantum obscure/fuzzy sets was considered. The so-called classical-quantum obscure/fuzzy registers were introduced in the first step (for $n = 2$, the minimal case) as

$$| s\rangle_f = \sqrt{1 - f}\, | 0\rangle + \sqrt{f}\, | 1\rangle, \tag{1.21}$$

$$| s\rangle_g = \sqrt{1 - g}\, | 0\rangle + \sqrt{g}\, | 1\rangle, \tag{1.22}$$

where $f, g \in [0, 1]$ are the relevant classical-quantum membership functions. In the second step their quantum superposition is defined by

$$| s\rangle = c_f\, | s\rangle_f + c_g\, | s\rangle_g, \tag{1.23}$$

where c_f and c_g are the probability amplitudes of the fuzzy states $| s\rangle_f$ and $| s\rangle_g$, respectively. It can be seen that the state (1.23) is a particular case of (1.8) with

$$\alpha_0 a_0 = c_f\sqrt{1 - f} + c_g\sqrt{1 - g}, \tag{1.24}$$

$$\alpha_1 a_1 = c_f\sqrt{f} + c_g\sqrt{g}. \tag{1.25}$$

This gives explicit connection of our double amplitude description of obscure qubits with the approach (Mannucci 2006, Maron *et al* 2013) which uses probability amplitudes and the membership functions. It is important to note that the use of the membership amplitudes introduced here α_i and (1.2) allows us to exploit the standard quantum gates but not to define new special ones, as in Mannucci (2006) and Maron *et al* (2013).

Another possible form of $F_{\mu_{0,\,1}}(\alpha_0, \alpha_1)$ (1.12), with the corresponding membership functions satisfying the standard fuzziness rules, can be found using a standard homeomorphism between the circle and the square. In Hannachi *et al* (2007b) and Rybalov *et al* (2014), this transformation was applied to the probability amplitudes $a_{0,\,1}$.

Innovation 1.3. *Here we exploit it for the membership amplitudes* $\alpha_{0,\,1}$

$$F_{\mu_0}(\alpha_0, \alpha_1) = \frac{2}{\pi}\arcsin\sqrt{\frac{\alpha_0^2 \operatorname{sign}\alpha_0 - \alpha_1^2 \operatorname{sign}\alpha_1 + 1}{2}}, \tag{1.26}$$

$$F_{\mu_1}(\alpha_0, \alpha_1) = \frac{2}{\pi}\arcsin\sqrt{\frac{\alpha_0^2 \operatorname{sign}\alpha_0 + \alpha_1^2 \operatorname{sign}\alpha_1 + 1}{2}}. \tag{1.27}$$

So for positive $\alpha_{0,\,1}$, we obtain (cf Hannachi *et al* 2007b)

$$F_{\mu_0}(\alpha_0, \alpha_1) = \frac{2}{\pi}\arcsin\sqrt{\frac{\alpha_0^2 - \alpha_1^2 + 1}{2}}, \tag{1.28}$$

$$F_{\mu_1}(\alpha_0, \alpha_1) = 1. \tag{1.29}$$

The equivalent membership functions for the outcome are

$$\max\left(\min\left(F_{\mu_0}(\alpha_0, \alpha_1), 1 - F_{\mu_1}(\alpha_0, \alpha_1)\right), \min\left(1 - F_{\mu_0}(\alpha_0, \alpha_1)\right), F_{\mu_1}(\alpha_0, \alpha_1)\right), \tag{1.30}$$

$$\min\left(\max\left(F_{\mu_0}(\alpha_0, \alpha_1), 1 - F_{\mu_1}(\alpha_0, \alpha_1)\right), \max\left(1 - F_{\mu_0}(\alpha_0, \alpha_1)\right), F_{\mu_1}(\alpha_0, \alpha_1)\right). \tag{1.31}$$

There are many different models for $F_{\mu_{0,1}}(\alpha_0, \alpha_1)$ which can be introduced in such a way that they satisfy the obscure set axioms (Dubois and Prade 2000, Zimmermann 2011).

1.3 Transformations of obscure qubits

Let us consider the obscure qubits in the vector representation, such that

$$|0\rangle = \begin{pmatrix} 1 \\ 0 \end{pmatrix}, \quad |1\rangle = \begin{pmatrix} 0 \\ 1 \end{pmatrix} \tag{1.32}$$

are basis vectors of the two-dimensional Hilbert space $\mathcal{H}_q^{(2)}$. A standard quantum computational process in the quantum register with L obscure qubits (qudits (1.1)) is performed by sequences of unitary matrices U of size $2^L \times 2^L$ ($n^L \times n^L$), $\mathsf{U}^\dagger \mathsf{U} = \mathsf{I}$, which are called quantum gates (I is the unit matrix). Thus, for one obscure qubit, the quantum gates are 2×2 unitary complex matrices.

Innovation 1.4. *In the vector representation, an obscure qubit differs from the standard qubit (1.8) by a 2×2 invertible diagonal (not necessarily unitary) matrix*

$$\left|\psi_{ob}^{(2)}\right\rangle = \mathsf{M}(\alpha_0, \alpha_1)\,|\psi^{(2)}\rangle, \tag{1.33}$$

$$\mathsf{M}(\alpha_0, \alpha_1) = \begin{pmatrix} \alpha_0 & 0 \\ 0 & \alpha_1 \end{pmatrix}. \tag{1.34}$$

We call $\mathsf{M}(\alpha_0, \alpha_1)$ a membership matrix which can optionally have the property

$$\mathrm{tr}\mathsf{M}^2 = 1, \tag{1.35}$$

if (1.15) holds.

Let us introduce the orthogonal commuting projection operators

$$\mathsf{P}_0 = \begin{pmatrix} 1 & 0 \\ 0 & 0 \end{pmatrix}, \quad \mathsf{P}_1 = \begin{pmatrix} 0 & 0 \\ 0 & 1 \end{pmatrix}, \tag{1.36}$$

$$\mathsf{P}_0^2 = \mathsf{P}_0, \ \ \mathsf{P}_1^2 = \mathsf{P}_1, \ \ \mathsf{P}_0\mathsf{P}_1 = \mathsf{P}_1\mathsf{P}_0 = \mathsf{0}, \tag{1.37}$$

where $\mathsf{0}$ is the 2×2 zero matrix. Well-known properties of the projections are that

$$\mathsf{P}_0 \mid \psi^{(2)}\rangle = a_0 \mid 0\rangle, \qquad \mathsf{P}_1 \mid \psi^{(2)}\rangle = a_1 \mid 0\rangle, \tag{1.38}$$

$$\langle\psi^{(2)} \mid \mathsf{P}_0 \mid \psi^{(2)}\rangle = \mid a_0 \mid^2, \qquad \langle\psi^{(2)} \mid \mathsf{P}_1 \mid \psi^{(2)}\rangle = \mid a_1 \mid^2. \tag{1.39}$$

Innovation 1.5. *The membership matrix (1.34) can be defined as a linear combination of the projection operators with the membership amplitudes as coefficients*

$$\mathsf{M}(\alpha_0, \alpha_1) = \alpha_0\mathsf{P}_0 + \alpha_1\mathsf{P}_1. \tag{1.40}$$

We compute

$$\mathsf{M}(\alpha_0, \alpha_1) \left| \psi_{\text{ob}}^{(2)} \right\rangle = \alpha_0^2 a_0 \mid 0\rangle + \alpha_1^2 a_1 \mid 1\rangle. \tag{1.41}$$

We can therefore treat the application of the membership matrix (1.33) as providing the origin of a reversible but non-unitary obscure measurement on the standard qubit to obtain an obscure qubit—cf the mirror measurement (Battilotti and Zizzi 2004, Zizzi 2005) and also the origin of ordinary qubit states on the fuzzy sphere (Zizzi and Pessa 2014).

An obscure analog of the density operator (for a pure state) is the following form for the density matrix in the vector representation

$$\rho_{\text{ob}}^{(2)} = \left| \psi_{\text{ob}}^{(2)} \right\rangle \left\langle \psi_{\text{ob}}^{(2)} \right| = \begin{pmatrix} \alpha_0^2 \mid a_0 \mid^2 & \alpha_0 a_0^* \alpha_1 a_1 \\ \alpha_0 a_0 \alpha_1 a_1^* & \alpha_1^2 \mid a_1 \mid^2 \end{pmatrix} \tag{1.42}$$

with the obvious standard singularity property $\det \rho_{\text{ob}}^{(2)} = 0$. But $\operatorname{tr}\rho_{\text{ob}}^{(2)} = \alpha_0^2 \mid a_0 \mid^2 + \alpha_1^2 \mid a_1 \mid^2 \neq 1$, and here there is no idempotence $\left(\rho_{\text{ob}}^{(2)}\right)^2 \neq \rho_{\text{ob}}^{(2)}$, which can distinct $\rho_{\text{ob}}^{(2)}$ from the standard density operator.

1.4 Kronecker obscure qubits

We next introduce an analog of quantum superposition for membership amplitudes, called 'obscure superposition' (cf Cunha *et al* 2019, and also Toffano and Dubois 2017).

Innovation 1.6. *Quantum amplitudes and membership amplitudes will here be considered separately in order to define an obscure qubit taking the form of a double superposition (cf (1.8), and a generalized analog for qudits (1.1) is straightforward)*

$$\mid \Psi_{\text{ob}}\rangle = \frac{\mathsf{A}_0 \mid 0\rangle + \mathsf{A}_1 \mid 1\rangle}{\sqrt{2}}, \tag{1.43}$$

where the two-dimensional vectors

$$\mathsf{A}_{0,1} = \begin{bmatrix} a_{0,1} \\ \alpha_{0,1} \end{bmatrix} \tag{1.44}$$

are the (double) obscure-quantum amplitudes of the generalized states $|0\rangle, |1\rangle$.

For the conjugate of an obscure qubit we put (informally)

$$\langle \Psi_{ob} | = \frac{\mathsf{A}_0^{\star}\langle 0| + \mathsf{A}_1^{\star}\langle 1|}{\sqrt{2}}, \tag{1.45}$$

where we denote $\mathsf{A}_{0,1}^{\star} = \begin{bmatrix} a_{0,1}^{*} & \alpha_{0,1} \end{bmatrix}$, such that $\mathsf{A}_{0,1}^{\star}\mathsf{A}_{0,1} = |a_{0,1}|^2 + \alpha_{0,1}^2$. The (double) obscure qubit is normalized in such a way that, if the conditions (1.14)–(1.15) hold, then

$$\langle \Psi_{ob} | \Psi_{ob} \rangle = \frac{|a_0|^2 + |a_1|^2}{2} + \frac{\alpha_0^2 + \alpha_1^2}{2} = 1. \tag{1.46}$$

Innovation 1.7. *A measurement should be made separately and independently in the probability space and the membership space, which can be represented using an analog of the Kronecker product.*

Indeed, in the vector representation (1.32) for the quantum states and for the direct product amplitudes (1.44) we should have

$$|\Psi_{ob}\rangle_{(0)} = \frac{1}{\sqrt{2}}\mathsf{A}_0 \otimes_K \begin{pmatrix} 1 \\ 0 \end{pmatrix} + \mathsf{A}_1 \otimes_K \begin{pmatrix} 0 \\ 1 \end{pmatrix}, \tag{1.47}$$

where the (left) Kronecker product is defined by (see (1.32))

$$\begin{bmatrix} a \\ \alpha \end{bmatrix} \otimes_K \begin{pmatrix} c \\ d \end{pmatrix} = \begin{bmatrix} a\begin{pmatrix} c \\ d \end{pmatrix} \\ \alpha\begin{pmatrix} c \\ d \end{pmatrix} \end{bmatrix} = \begin{bmatrix} a(c\mathsf{e}_0 + d\mathsf{e}_1) \\ \alpha(c\mathsf{e}_0 + d\mathsf{e}_1) \end{bmatrix},$$
$$\mathsf{e}_0 = \begin{pmatrix} 1 \\ 0 \end{pmatrix}, \ \mathsf{e}_1 = \begin{pmatrix} 0 \\ 1 \end{pmatrix}, \ \mathsf{e}_{0,1} \in \mathcal{H}_q^{(2)}. \tag{1.48}$$

Informally, the wave function of the obscure qubit, in the vector representation, now lives in the four-dimensional space of (1.48), which has two two-dimensional spaces as blocks. The upper block, the quantum subspace, is the ordinary Hilbert space $\mathcal{H}_q^{(2)}$, but the lower block should have special (fuzzy) properties, if it is treated as an obscure (membership) subspace $\mathcal{V}_{memb}^{(2)}$. Thus, the four-dimensional space, where lives $\left|\Psi_{ob}^{(2)}\right\rangle$, is not an ordinary tensor product of vector spaces because of (1.48) and the vector A on the lhs has entries of different natures, i.e., the quantum amplitudes $a_{0,\,1}$ and the membership amplitudes $\alpha_{0,\,1}$. Despite the unit vectors in $\mathcal{H}_{\mathrm{q}}^{(2)}$ and $\mathcal{V}_{\mathrm{memb}}^{(2)}$ having the same form (1.32), they belong to different spaces (because they are vector spaces over different fields). Therefore, instead of (1.48), we introduce a Kronecker-like product $\tilde{\otimes}_{\mathrm{K}}$ by

$$\begin{bmatrix} a \\ \alpha \end{bmatrix} \tilde{\otimes}_{\mathrm{K}} \begin{pmatrix} c \\ d \end{pmatrix} = \begin{bmatrix} a(c\mathsf{e}_0 + d\mathsf{e}_1) \\ \alpha(c\varepsilon_0 + d\varepsilon_1) \end{bmatrix}, \tag{1.49}$$

$$\mathsf{e}_0 = \begin{pmatrix} 1 \\ 0 \end{pmatrix}, \quad \mathsf{e}_1 = \begin{pmatrix} 0 \\ 1 \end{pmatrix}, \quad \mathsf{e}_{0,\,1} \in \mathcal{H}_{\mathrm{q}}^{(2)}, \tag{1.50}$$

$$\varepsilon_0 = \begin{pmatrix} 1 \\ 0 \end{pmatrix}^{(\mu)}, \quad \varepsilon_1 = \begin{pmatrix} 0 \\ 1 \end{pmatrix}^{(\mu)}, \quad \varepsilon_{0,\,1} \in \mathcal{V}_{\mathrm{memb}}^{(2)}. \tag{1.51}$$

In this way, the obscure qubit (1.43) can be presented in the from

$$\begin{aligned} | \mathbf{\Psi}_{\mathrm{ob}} \rangle &= \frac{1}{\sqrt{2}} \begin{bmatrix} a_0 \begin{pmatrix} 1 \\ 0 \end{pmatrix} \\ \alpha_0 \begin{pmatrix} 1 \\ 0 \end{pmatrix}^{(\mu)} \end{bmatrix} + \frac{1}{\sqrt{2}} \begin{bmatrix} a_1 \begin{pmatrix} 0 \\ 1 \end{pmatrix} \\ \alpha_1 \begin{pmatrix} 0 \\ 1 \end{pmatrix}^{(\mu)} \end{bmatrix} \\ &= \frac{1}{\sqrt{2}} \begin{bmatrix} a_0 \mathsf{e}_0 \\ \alpha_0 \varepsilon_0 \end{bmatrix} + \frac{1}{\sqrt{2}} \begin{bmatrix} a_1 \mathsf{e}_1 \\ \alpha_1 \varepsilon_1 \end{bmatrix}. \end{aligned} \tag{1.52}$$

Therefore, we call the double obscure qubit (1.52) a Kronecker obscure qubit to distinguish it from the obscure qubit (1.8). It can be also presented using the Hadamard product (the element-wise or Schur product)

$$\begin{bmatrix} a \\ \alpha \end{bmatrix} \otimes_{\mathrm{H}} \begin{pmatrix} c \\ d \end{pmatrix} = \begin{bmatrix} ac \\ \alpha d \end{bmatrix} \tag{1.53}$$

in the following form

$$| \mathbf{\Psi}_{\mathrm{ob}} \rangle = \frac{1}{\sqrt{2}} \mathsf{A}_0 \otimes_{\mathrm{H}} \mathsf{E}_0 + \frac{1}{\sqrt{2}} \mathsf{A}_1 \otimes_{\mathrm{H}} \mathsf{E}_1, \tag{1.54}$$

where the unit vectors of the total four-dimensional space are

$$\mathsf{E}_{0,\,1} = \begin{bmatrix} \mathsf{e}_{0,\,1} \\ \varepsilon_{0,\,1} \end{bmatrix} \in \mathcal{H}_{\mathrm{q}}^{(2)} \times \mathcal{V}_{\mathrm{memb}}^{(2)}. \tag{1.55}$$

The probabilities $p_{0,\,1}$ and membership functions $\mu_{0,\,1}$ of the states $|\,0\rangle$ and $|\,1\rangle$ are computed through the corresponding amplitudes by (1.11) and (1.12)

$$p_i = |\, a_i \,|^2, \quad \mu_i = F_{\mu_i}(\alpha_0, \alpha_1), \quad i = 0, 1, \tag{1.56}$$

and in the particular case by (1.13) satisfying (1.15).

By way of example, consider a Kronecker obscure qubit (with a real quantum part) with probability p and membership function μ (measure of trust) of the state $|\,0\rangle$, and of the state $|\,1\rangle$ given by $1-p$ and $1-\mu$, respectively. In the model (1.19)–(1.20) for μ_i (which is not Born-like) we obtain

$$|\,\Psi_{\text{ob}}\rangle = \frac{1}{\sqrt{2}}\begin{bmatrix}\begin{pmatrix}\sqrt{p}\\0\end{pmatrix}\\ \begin{pmatrix}\cos\frac{\pi}{2}\mu\\0\end{pmatrix}^{(\mu)}\end{bmatrix} + \frac{1}{\sqrt{2}}\begin{bmatrix}\begin{pmatrix}0\\\sqrt{1-p}\end{pmatrix}\\ \begin{pmatrix}0\\\sin\frac{\pi}{2}\mu\end{pmatrix}^{(\mu)}\end{bmatrix} \tag{1.57}$$
$$= \frac{1}{\sqrt{2}}\begin{bmatrix}\mathsf{e}_0\sqrt{p}\\ \varepsilon_0\cos\frac{\pi}{2}\mu\end{bmatrix} + \frac{1}{\sqrt{2}}\begin{bmatrix}\mathsf{e}_1\sqrt{1-p}\\ \varepsilon_1\sin\frac{\pi}{2}\mu\end{bmatrix},$$

where e_i and ε_i are unit vectors defined in (1.50) and (1.51).

This can be compared, e.g., with the classical-quantum approach (1.23), and Mannucci (2006) and Maron *et al* (2013), in which the elements of the columns are multiplied, while we consider them independently and separately.

1.5 Obscure-quantum measurement

Let us consider the case of one Kronecker obscure qubit register $L = 1$ (see (1.47)), or using (1.48) in the vector representation (1.52). The standard (double) orthogonal commuting projection operators, Kronecker projections, are (cf (1.36))

$$\mathbf{P}_0 = \begin{bmatrix}\mathsf{P}_0 & 0\\ 0 & \mathsf{P}_0^{(\mu)}\end{bmatrix},\quad \mathbf{P}_1 = \begin{bmatrix}\mathsf{P}_1 & 0\\ 0 & \mathsf{P}_1^{(\mu)}\end{bmatrix}, \tag{1.58}$$

where $\mathsf{0}$ is the 2×2 zero matrix, and $\mathsf{P}_{0,1}^{(\mu)}$ are the projections in the membership subspace $\mathcal{V}_{\text{memb}}^{(2)}$ (of the same form as the ordinary quantum projections $\mathsf{P}_{0,1}$ (1.36))

$$\mathsf{P}_0^{(\mu)} = \begin{pmatrix}1 & 0\\ 0 & 0\end{pmatrix}^{(\mu)},\quad \mathsf{P}_1^{(\mu)} = \begin{pmatrix}0 & 0\\ 0 & 1\end{pmatrix}^{(\mu)},\quad \mathsf{P}_0^{(\mu)}, \mathsf{P}_1^{(\mu)} \in \operatorname{End}\mathcal{V}_{\text{memb}}^{(2)}, \tag{1.59}$$

$$\mathsf{P}_0^{(\mu)2} = \mathsf{P}_0^{(\mu)},\ \mathsf{P}_1^{(\mu)2} = \mathsf{P}_1^{(\mu)},\ \mathsf{P}_0^{(\mu)}\mathsf{P}_1^{(\mu)} = \mathsf{P}_1^{(\mu)}\mathsf{P}_0^{(\mu)} = \mathsf{0}. \tag{1.60}$$

For the double projections we have (cf (1.37))

$$\mathbf{P}_0^2 = \mathbf{P}_0,\ \mathbf{P}_1^2 = \mathbf{P}_1,\ \mathbf{P}_0\mathbf{P}_1 = \mathbf{P}_1\mathbf{P}_0 = \mathbf{0}, \tag{1.61}$$

where $\mathbf{0}$ is the 4×4 zero matrix, and $\mathbf{P}_{0,1}$ act on the Kronecker qubit (1.58) in the standard way (cf (1.38))

$$\mathbf{P}_0 \mid \boldsymbol{\Psi}_{\text{ob}}\rangle = \frac{1}{\sqrt{2}}\begin{bmatrix} \alpha_0\begin{pmatrix}1\\0\end{pmatrix} \\ \alpha_0\begin{pmatrix}1\\0\end{pmatrix}^{(\mu)} \end{bmatrix} = \frac{1}{\sqrt{2}}\begin{bmatrix} a_0\mathsf{e}_0 \\ \alpha_0\varepsilon_0 \end{bmatrix} = \frac{1}{\sqrt{2}}\mathsf{A}_0 \otimes_{\text{H}} \mathsf{E}_0, \tag{1.62}$$

$$\mathbf{P}_1 \mid \boldsymbol{\Psi}_{\text{ob}}\rangle = \frac{1}{\sqrt{2}}\begin{bmatrix} a_1\begin{pmatrix}0\\1\end{pmatrix} \\ \alpha_1\begin{pmatrix}0\\1\end{pmatrix}^{(\mu)} \end{bmatrix} = \frac{1}{\sqrt{2}}\begin{bmatrix} a_1\mathsf{e}_1 \\ \alpha_1\varepsilon_1 \end{bmatrix} = \frac{1}{\sqrt{2}}\mathsf{A}_1 \otimes_{\text{H}} \mathsf{E}_1. \tag{1.63}$$

Observe that for Kronecker qubits there exist in addition to (1.58) the following orthogonal commuting projection operators

$$\mathbf{P}_{01} = \begin{bmatrix} \mathsf{P}_0 & 0 \\ 0 & \mathsf{P}_1^{(\mu)} \end{bmatrix}, \quad \mathbf{P}_{10} = \begin{bmatrix} \mathsf{P}_1 & 0 \\ 0 & \mathsf{P}_0^{(\mu)} \end{bmatrix}, \tag{1.64}$$

and we call these the crossed double projections. They satisfy the same relations as (1.61)

$$\mathbf{P}_{01}^2 = \mathbf{P}_{01}, \;\; \mathbf{P}_{10}^2 = \mathbf{P}_{10}, \;\; \mathbf{P}_{01}\mathbf{P}_{10} = \mathbf{P}_{10}\mathbf{P}_{01} = \mathbf{0}, \tag{1.65}$$

but act on the obscure qubit in a different (mixing) way than (1.62), i.e.,

$$\mathbf{P}_{01} \mid \boldsymbol{\Psi}_{\text{ob}}\rangle = \frac{1}{\sqrt{2}}\begin{bmatrix} a_0\begin{pmatrix}1\\0\end{pmatrix} \\ \alpha_1\begin{pmatrix}0\\1\end{pmatrix} \end{bmatrix} = \frac{1}{\sqrt{2}}\begin{bmatrix} a_0\mathsf{e}_0 \\ \alpha_1\varepsilon_1 \end{bmatrix}, \tag{1.66}$$

$$\mathbf{P}_{10} \mid \boldsymbol{\Psi}_{\text{ob}}\rangle = \frac{1}{\sqrt{2}}\begin{bmatrix} a_1\begin{pmatrix}0\\1\end{pmatrix} \\ \alpha_0\begin{pmatrix}1\\0\end{pmatrix} \end{bmatrix} = \frac{1}{\sqrt{2}}\begin{bmatrix} a_1\mathsf{e}_1 \\ \alpha_0\varepsilon_0 \end{bmatrix}. \tag{1.67}$$

The multiplication of the crossed double projections (1.64) and the double projections (1.58) is given by

$$\mathbf{P}_{01}\mathbf{P}_0 = \mathbf{P}_0\mathbf{P}_{01} = \begin{bmatrix} \mathsf{P}_0 & 0 \\ 0 & 0 \end{bmatrix} \equiv \mathbf{Q}_0, \quad \mathbf{P}_{01}\mathbf{P}_1 = \mathbf{P}_1\mathbf{P}_{01} = \begin{bmatrix} 0 & 0 \\ 0 & \mathsf{P}_1^{(\mu)} \end{bmatrix} \equiv \mathbf{Q}_1^{(\mu)}, \tag{1.68}$$

$$\mathbf{P}_{10}\mathbf{P}_0 = \mathbf{P}_0\mathbf{P}_{10} = \begin{bmatrix} 0 & 0 \\ 0 & \mathsf{P}_0^{(\mu)} \end{bmatrix} \equiv \mathbf{Q}_0^{(\mu)}, \quad \mathbf{P}_{10}\mathbf{P}_1 = \mathbf{P}_1\mathbf{P}_{10} = \begin{bmatrix} \mathsf{P}_1 & 0 \\ 0 & 0 \end{bmatrix} \equiv \mathbf{Q}_1, \tag{1.69}$$

where the operators $\mathbf{Q}_0$, $\mathbf{Q}_1$ and $\mathbf{Q}_0^{(\mu)}$, $\mathbf{Q}_1^{(\mu)}$ satisfy

$$\mathbf{Q}_0^2 = \mathbf{Q}_0,\ \ \mathbf{Q}_1^2 = \mathbf{Q}_1,\ \ \mathbf{Q}_1\mathbf{Q}_0 = \mathbf{Q}_0\mathbf{Q}_1 = \mathbf{0}, \tag{1.70}$$

$$\mathbf{Q}_0^{(\mu)2} = \mathbf{Q}_0^{(\mu)},\ \ \mathbf{Q}_1^{(\mu)2} = \mathbf{Q}_1^{(\mu)},\ \ \mathbf{Q}_1^{(\mu)}\mathbf{Q}_0^{(\mu)} = \mathbf{Q}_0^{(\mu)}\mathbf{Q}_1^{(\mu)} = \mathbf{0}, \tag{1.71}$$

$$\mathbf{Q}_1^{(\mu)}\mathbf{Q}_0 = \mathbf{Q}_0^{(\mu)}\mathbf{Q}_1 = \mathbf{Q}_1\mathbf{Q}_0^{(\mu)} = \mathbf{Q}_0\mathbf{Q}_1^{(\mu)} = \mathbf{0}, \tag{1.72}$$

and we call these ‘half Kronecker (double) projections’.

These relations imply that the process of measurement when using Kronecker obscure qubits (i.e. for quantum computation with truth or membership) is more complicated than in the standard case.

To show this, let us calculate the obscure analogs of expected values for the projections above. Using the notation

$$\bar{\mathbf{A}} \equiv \langle \mathbf{\Psi}_{\mathrm{ob}} \mid \mathbf{A} \mid \mathbf{\Psi}_{\mathrm{ob}}\rangle. \tag{1.73}$$

Then, using (1.43)–(1.45) for the projection operators $\mathbf{P}_i$, $\mathbf{P}_{ij}$, $\mathbf{Q}_i$, $\mathbf{Q}_i^{(\mu)}$, $i, j = 0, 1$, $i \neq j$, we obtain (cf (1.39))

$$\bar{\mathbf{P}}_i = \frac{|a_i|^2 + \alpha_i^2}{2},\quad \bar{\mathbf{P}}_{ij} = \frac{|a_i|^2 + \alpha_j^2}{2}, \tag{1.74}$$

$$\bar{\mathbf{Q}}_i = \frac{|a_i|^2}{2},\qquad \bar{\mathbf{Q}}_i^{(\mu)} = \frac{\alpha_i^2}{2}. \tag{1.75}$$

So follows the relation between the obscure analogs of expected values of the projections

$$\bar{\mathbf{P}}_i = \bar{\mathbf{Q}}_i + \bar{\mathbf{Q}}_i^{(\mu)},\quad \bar{\mathbf{P}}_{ij} = \bar{\mathbf{Q}}_i + \bar{\mathbf{Q}}_j^{(\mu)}. \tag{1.76}$$

Taking the ket corresponding to the bra Kronecker qubit (1.52) in the form

$$\langle \mathbf{\Psi}_{\mathrm{ob}} \mid = \frac{1}{\sqrt{2}}\left[a_0^*(1\ \ 0),\ \alpha_0(1\ \ 0)\right] + \frac{1}{\sqrt{2}}\left[a_1^*(0\ \ 1),\ \alpha_1(0\ \ 1)\right], \tag{1.77}$$

a Kronecker (4 × 4) obscure analog of the density matrix for a pure state is given by (cf (1.42))

$$\rho_{\mathrm{ob}}^{(2)} = |\,\mathbf{\Psi}_{\mathrm{ob}}\rangle\langle\mathbf{\Psi}_{\mathrm{ob}}\,| = \frac{1}{2}\begin{pmatrix} |a_0|^2 & a_0a_1^* & a_0\alpha_0 & a_0\alpha_1 \\ a_1a_0^* & |a_1|^2 & a_1\alpha_0 & a_1\alpha_1 \\ \alpha_0a_0^* & \alpha_0a_1^* & \alpha_0^2 & \alpha_0\alpha_1 \\ \alpha_1a_0^* & \alpha_1a_1^* & \alpha_0\alpha_1 & \alpha_1^2 \end{pmatrix}. \tag{1.78}$$

If the Born rule for the membership functions (1.13) and the conditions (1.14)–(1.15) are satisfied, then the density matrix (1.78) is non-invertible because $\det \rho_{\mathrm{ob}}^{(2)} = 0$ and has unit trace $\mathrm{tr}\rho_{\mathrm{ob}}^{(2)} = 1$ but is not idempotent $\left(\rho_{\mathrm{ob}}^{(2)}\right)^2 \neq \rho_{\mathrm{ob}}^{(2)}$ because it holds for the ordinary quantum density matrix (Nielsen and Chuang 2000).

1.6 Kronecker obscure-quantum gates

In general, (double) obscure-quantum computation with L Kronecker obscure qubits (or qudits) can be performed by a product of unitary (block) matrices $\mathbf{U}$ of the (double size to the standard one) size $2 \times (2^L \times 2^L)$ (or $2 \times (n^L \times n^L)$), $\mathbf{U}^{\dagger}\mathbf{U} = \mathbf{I}$ (here $\mathbf{I}$ is the unit matrix of the same size as $\mathbf{U}$). We can also call such computation a quantum computation with truth (or with membership).

Let us consider obscure-quantum computation with one Kronecker obscure qubit. Informally, we can present the Kronecker obscure qubit (1.52) in the form

$$|\,\Psi_{\text{ob}}\rangle = \begin{bmatrix} \frac{1}{\sqrt{2}}\begin{pmatrix} a_0 \\ a_1 \end{pmatrix} \\ \frac{1}{\sqrt{2}}\begin{pmatrix} \alpha_0 \\ \alpha_1 \end{pmatrix}^{(\mu)} \end{bmatrix}. \tag{1.79}$$

Innovation 1.8. *The state* $|\,\Psi_{\text{ob}}\rangle$ *can be interpreted as a vector in the direct product (not tensor product) space* $\mathcal{H}_{\text{q}}^{(2)} \times \mathcal{V}_{\text{memb}}^{(2)}$, *where* $\mathcal{H}_{\text{q}}^{(2)}$ *is the standard two-dimensional Hilbert space of the qubit, and* $\mathcal{V}_{\text{memb}}^{(2)}$ *can be treated as the membership space, which has a different nature from the qubit space and can have a more complex structure.*

For discussion of similar spaces, see for example Dubois *et al* (2000), Bělohlávek (2002), Smith (2008), and Zimmermann (2011). In general, one can consider obscure-quantum computation as a set of abstract computational rules, independently of the introduction of the corresponding spaces.

An obscure-quantum gate will be defined as an elementary transformation on an obscure qubit (1.79) and is performed by unitary (block) matrices of size 4×4 (over $\mathbb{C}$) acting in the total space $\mathcal{H}_{\text{q}}^{(2)} \times \mathcal{V}_{\text{memb}}^{(2)}$

$$\mathbf{U} = \begin{pmatrix} \mathsf{U} & 0 \\ 0 & \mathsf{U}^{(\mu)} \end{pmatrix}, \quad \mathbf{U}\mathbf{U}^{\dagger} = \mathbf{U}^{\dagger}\mathbf{U} = \mathbf{I}, \tag{1.80}$$

$$\mathsf{U}\mathsf{U}^{\dagger} = \mathsf{U}^{\dagger}\mathsf{U} = \mathsf{I}, \quad \mathsf{U}^{(\mu)}\mathsf{U}^{(\mu)\dagger} = \mathsf{U}^{(\mu)\dagger}\mathsf{U}^{(\mu)} = \mathsf{I}, \quad \mathsf{U} \in \operatorname{End}\mathcal{H}_{\text{q}}^{(2)}, \, \mathsf{U}^{(\mu)} \in \operatorname{End}\mathcal{V}_{\text{memb}}^{(2)}, \tag{1.81}$$

where $\mathbf{I}$ is the unit 4×4 matrix, I is the unit 2×2 matrix, and U and $\mathsf{U}^{(\mu)}$ are unitary 2×2 matrices acting on the probability and membership subspaces, respectively. The matrix U (over $\mathbb{C}$) will be called a quantum gate, and we call the matrix $\mathsf{U}^{(\mu)}$ (over $\mathbb{R}$) an obscure gate. We assume that the obscure gates $\mathsf{U}^{(\mu)}$ are of the same shape as the standard quantum gates, but they act in the other (membership) space and have only real elements (see, e.g. Nielsen and Chuang 2000). In this case, an obscure-quantum gate is characterized by the pair $\{\mathsf{U}, \mathsf{U}^{(\mu)}\}$, where the components are known gates (in various combinations), e.g., for one qubit gates: Hadamard, Pauli-X (`NOT`),Y,Z (or two qubit gates e.g. `CNOT`, `SWAP`, etc). The transformed qubit then becomes (informally)

$$\mathbf{U} \mid \Psi_{ob}\rangle = \begin{bmatrix} \frac{1}{\sqrt{2}} \mathsf{U} \begin{pmatrix} a_0 \\ a_1 \end{pmatrix} \\ \frac{1}{\sqrt{2}} \mathsf{U}^{(\mu)} \begin{pmatrix} \alpha_0 \\ \alpha_1 \end{pmatrix}^{(\mu)} \end{bmatrix}. \tag{1.82}$$

Thus, the quantum and the membership parts are transformed independently for the block diagonal form (1.80). Some examples of this can be found, e.g., in Domenech and Freytes (2006), Mannucci (2006), and Maron *et al* (2013). Differences between the parts were mentioned in Kreinovich *et al* (2011). In this case, an obscure-quantum network is physically realized by a device performing elementary operations in sequence on obscure qubits (by a product of matrices), such that the quantum and membership parts are synchronized in time; for a discussion of the obscure part of such physical devices, see Hirota and Ozawa (1989), Kóczy and Hirota (1990), Virant (2000), and Kosko (1997). Then, the result of the obscure-quantum computation consists of the quantum probabilities of the states together with the calculated level of truth for each of them (see, e.g. Bolotin 2018).

For example, the obscure-quantum gate $\mathbf{U}_{H,NOT} = \{\text{Hadamard, NOT}\}$ acts on the state E_0 (1.55) as follows

$$\mathbf{U}_{H,NOT}\mathsf{E}_0 = \mathbf{U}_{H,NOT} \begin{bmatrix} \begin{pmatrix} 1 \\ 0 \end{pmatrix} \\ \begin{pmatrix} 1 \\ 0 \end{pmatrix}^{(\mu)} \end{bmatrix} = \begin{bmatrix} \frac{1}{\sqrt{2}} \begin{pmatrix} 1 \\ 1 \end{pmatrix} \\ \begin{pmatrix} 0 \\ 1 \end{pmatrix}^{(\mu)} \end{bmatrix} = \begin{bmatrix} \frac{1}{\sqrt{2}} (\mathsf{e}_0 + \mathsf{e}_1) \\ \varepsilon_1 \end{bmatrix}. \tag{1.83}$$

It would be interesting to consider the case when $\mathbf{U}$ (1.80) is not block diagonal and try to find possible physical interpretations of the non-diagonal blocks.

1.7 Double entanglement

Let us introduce a register consisting of two obscure qubits ($L = 2$) in the computational basis $\mid ij'\rangle = \mid i\rangle \otimes \mid j'\rangle$, as follows

$$\left| \Psi_{ob}^{(n=2)}(L = 2) \right\rangle = \mid \Psi_{ob}(2)\rangle = \frac{\mathsf{B}_{00'} \mid 00'\rangle + \mathsf{B}_{10'} \mid 10'\rangle + \mathsf{B}_{01'} \mid 01'\rangle + \mathsf{B}_{11'} \mid 11'\rangle}{\sqrt{2}}, \tag{1.84}$$

determined by two-dimensional vectors (encoding obscure-quantum amplitudes)

$$\mathsf{B}_{ij'} = \begin{bmatrix} b_{ij'} \\ \beta_{ij'} \end{bmatrix}, \quad i, j = 0, 1, \quad j' = 0', 1', \tag{1.85}$$

where $b_{ij'} \in \mathbb{C}$ are probability amplitudes for a set of pure states and $\beta_{ij'} \in \mathbb{R}$ are the corresponding membership amplitudes. By analogy with (1.43) and (1.46), the normalization factor in (1.84) is chosen so that

$$\langle \Psi_{ob}(2) | \Psi_{ob}(2)\rangle = 1, \tag{1.86}$$

if (cf (1.14)–(1.15))

$$|b_{00'}|^2 + |b_{10'}|^2 + |b_{01'}|^2 + |b_{11'}|^2 = 1, \tag{1.87}$$

$$\beta_{00'}^2 + \beta_{10'}^2 + \beta_{01'}^2 + \beta_{11'}^2 = 1. \tag{1.88}$$

A state of two qubits is entangled if it cannot be decomposed as a product of two one-qubit states, and otherwise it is separable (see, e.g. Nielsen and Chuang 2000).

Innovation 1.9. *We define a product of two obscure qubits (1.43) as*

$$|\Psi_{ob}\rangle \otimes |\Psi'_{ob}\rangle = \frac{A_0 \otimes_H A'_0 |00'\rangle + A_1 \otimes_H A'_0 |10'\rangle + A_0 \otimes_H A'_1 |01'\rangle + A_1 \otimes_H A'_1 |11'\rangle}{2}, \tag{1.89}$$

where $\otimes_H$ *is the Hadamard product (1.53).*

Comparing (1.84) and (1.89), we obtain two sets of relations, for probability amplitudes and for membership amplitudes

$$b_{ij'} = \frac{1}{\sqrt{2}} a_i a_{j'}, \tag{1.90}$$

$$\beta_{ij'} = \frac{1}{\sqrt{2}} \alpha_i \alpha_{j'}, \quad i, j = 0, 1, \quad j' = 0', 1'. \tag{1.91}$$

In this case, the relations (1.14)–(1.15) give (1.87)–(1.88).

Two obscure-quantum qubits are entangled if their joint state (1.84) cannot be presented as a product of one qubit states (1.89), and in the opposite case the states are called totally separable. It follows from (1.90)–(1.91) that there are two general conditions for obscure qubits to be entangled

$$b_{00'} b_{11'} \neq b_{10'} b_{01'}, \quad \text{or } \det \mathbf{b} \neq 0, \; \mathbf{b} = \begin{pmatrix} b_{00'} & b_{01'} \\ b_{10'} & b_{11'} \end{pmatrix}, \tag{1.92}$$

$$\beta_{00'} \beta_{11'} \neq \beta_{10'} \beta_{01'}, \quad \text{or } \det \boldsymbol{\beta} \neq 0, \; \boldsymbol{\beta} = \begin{pmatrix} \beta_{00'} & \beta_{01'} \\ \beta_{10'} & \beta_{11'} \end{pmatrix}. \tag{1.93}$$

The first equation (1.92) is the entanglement relation for the standard qubit, while the second condition (1.93) is for the membership amplitudes of the two obscure qubit joint state (1.84). The presence of two different conditions (1.92)–(1.93) leads to new additional possibilities (which do not exist for ordinary qubits) for partial entanglement (or partial separability), when only one of them is fulfilled. In this case, the states can be entangled in one subspace (quantum or membership) but not in the other.

The measure of entanglement is numerically characterized by the concurrence. Taking into account the two conditions (1.92)–(1.93), we propose to generalize the notion of concurrence for two obscure qubits in two ways. First, we introduce the vector obscure concurrence

$$\mathsf{C}_{\text{vect}} = \begin{bmatrix} C_{\text{q}} \\ C^{(\mu)} \end{bmatrix} = 2\begin{bmatrix} |\det \mathbf{b}| \\ |\det \beta| \end{bmatrix}, \tag{1.94}$$

where $\mathbf{b}$ and β are defined in (1.92)–(1.93), and $0 \leqslant C_{\text{q}} \leqslant 1$, $0 \leqslant C^{(\mu)} \leqslant 1$.

Innovation 1.10. *The corresponding scalar obscure concurrence can be defined as*

$$C_{\text{scal}} = \sqrt{\frac{|\det \mathbf{b}|^2 + |\det \beta|^2}{2}}, \tag{1.95}$$

such that $0 \leqslant C_{\text{scal}} \leqslant 1$. *Thus, two obscure qubits are totally separable, if* $C_{\text{scal}} = 0$.

For instance, for an obscure analog of the (maximally entangled) Bell state

$$|\Psi_{\text{ob}}(2)\rangle = \frac{1}{\sqrt{2}}\left(\begin{bmatrix} \frac{1}{\sqrt{2}} \\ \frac{1}{\sqrt{2}} \end{bmatrix}|00'\rangle + \begin{bmatrix} \frac{1}{\sqrt{2}} \\ \frac{1}{\sqrt{2}} \end{bmatrix}|11'\rangle\right) \tag{1.96}$$

we obtain

$$\mathsf{C}_{\text{vect}} = \begin{bmatrix} 1 \\ 1 \end{bmatrix}, \quad C_{\text{scal}} = 1. \tag{1.97}$$

A more interesting example is the intermediately entangled two obscure qubit state, e.g.,

$$|\Psi_{\text{ob}}(2)\rangle = \frac{1}{\sqrt{2}}\left(\begin{bmatrix} \frac{1}{2} \\ \frac{1}{\sqrt{2}} \end{bmatrix}|00'\rangle + \begin{bmatrix} \frac{1}{4} \\ \frac{\sqrt{5}}{4} \end{bmatrix}|10'\rangle + \begin{bmatrix} \frac{\sqrt{3}}{4} \\ \frac{1}{2\sqrt{2}} \end{bmatrix}|01'\rangle + \begin{bmatrix} \frac{1}{\sqrt{2}} \\ \frac{1}{4} \end{bmatrix}|11'\rangle\right), \tag{1.98}$$

where the amplitudes satisfy (1.87)–(1.88). If the Born-like rule (as in (1.13)) holds for the membership amplitudes, then the probabilities and membership functions of the states in (1.98) are

$$p_{00'} = \frac{1}{4}, \; p_{10'} = \frac{1}{16}, \; p_{01'} = \frac{3}{16}, \; p_{11'} = \frac{1}{2}, \tag{1.99}$$

$$\mu_{00'} = \frac{1}{2}, \; \mu_{10'} = \frac{5}{16}, \; \mu_{01'} = \frac{1}{8}, \; \mu_{11'} = \frac{1}{16}. \tag{1.100}$$

This means that, e.g., that the state $|10'\rangle$ will be measured with the quantum probability 1/16 and the membership function (truth value) 5/16. For the entangled obscure qubit (1.98) we obtain the concurrences

$$\mathsf{C}_{\text{vect}} = \begin{bmatrix} \frac{1}{2}\sqrt{2} - \frac{1}{8}\sqrt{3} \\ \frac{1}{8}\sqrt{2}\sqrt{5} - \frac{1}{4}\sqrt{2} \end{bmatrix} = \begin{bmatrix} 0.491 \\ 0.042 \end{bmatrix},$$

$$C_{\text{scal}} = \sqrt{\frac{53}{128} - \frac{1}{16}\sqrt{5} - \frac{1}{16}\sqrt{2}\sqrt{3}} = 0.348. \tag{1.101}$$

In the vector representation (1.49)–(1.52), we have

$$|\,\mathsf{ij}'\rangle = |\,\mathsf{i}\rangle \otimes |\,\mathsf{j}'\rangle = \begin{bmatrix} \mathsf{e}_i \otimes_{\mathrm{K}} \mathsf{e}_{j'} \\ \varepsilon_i \otimes_{\mathrm{K}} \varepsilon_{j'} \end{bmatrix}, \quad i, j = 0, 1, \quad j' = 0', 1', \tag{1.102}$$

where $\otimes_{\mathrm{K}}$ is the Kronecker product (1.48), and e_i, ε_i are defined in (1.50)–(1.51). Using (1.85) and the Kronecker-like product (1.49), we put (informally, with no summation)

$$\mathsf{B}_{ij'}\,|\,\mathsf{ij}'\rangle = \begin{bmatrix} b_{ij'}\mathsf{e}_i \otimes_{\mathrm{K}} \mathsf{e}_{j'} \\ \beta_{ij'}\varepsilon_i \otimes_{\mathrm{K}} \varepsilon_{j'} \end{bmatrix}, \quad i, j = 0, 1, \quad j' = 0', 1'. \tag{1.103}$$

To clarify our model, we show here a manifest form of the two obscure qubit state (1.98) in the vector representation

$$|\,\Psi_{\text{ob}}(2)\rangle = \frac{1}{\sqrt{2}}\left(\begin{bmatrix} \frac{1}{2}\begin{pmatrix}1\\0\\1\\0\end{pmatrix} \\ \frac{1}{\sqrt{2}}\begin{pmatrix}1\\0\\1\\0\end{pmatrix}^{(\mu)} \end{bmatrix} + \begin{bmatrix} \frac{1}{4}\begin{pmatrix}0\\1\\1\\0\end{pmatrix} \\ \frac{\sqrt{5}}{4}\begin{pmatrix}0\\1\\1\\0\end{pmatrix}^{(\mu)} \end{bmatrix} + \begin{bmatrix} \frac{\sqrt{3}}{4}\begin{pmatrix}1\\0\\0\\1\end{pmatrix} \\ \frac{1}{2\sqrt{2}}\begin{pmatrix}1\\0\\0\\1\end{pmatrix}^{(\mu)} \end{bmatrix} + \begin{bmatrix} \frac{1}{\sqrt{2}}\begin{pmatrix}0\\1\\0\\1\end{pmatrix} \\ \frac{1}{4}\begin{pmatrix}0\\1\\0\\1\end{pmatrix}^{(\mu)} \end{bmatrix} \right). \tag{1.104}$$

Innovation 1.11. *The states above may be called 'symmetric two obscure qubit states'. However, there are more general possibilities, as may be seen from the rhs of (1.103) and (1.104), when the indices of the first and second rows do not coincide. This would allow more possible states, which we call 'non-symmetric two obscure qubit states'. It would be worthwhile to establish their possible physical interpretation.*

These constructions show that quantum computing using Kronecker obscure qubits can involve a rich structure of states, giving a more detailed description with additional variables reflecting vagueness.

1.8 Conclusions

We have proposed a new scheme for describing quantum computation bringing vagueness into consideration, in which each state is characterized by a measure of truth. A membership amplitude is introduced in addition to the probability amplitude in order to achieve this, and we are led thereby to the concept of an obscure qubit. Two kinds of these are considered: the product obscure qubit, in which the total amplitude is the product of the quantum and membership amplitudes; and the Kronecker obscure qubit, where the amplitudes are manipulated separately. In the latter case, the quantum part of the computation is based, as usual, in Hilbert space, while the truth part requires a vague/fuzzy set formalism, which can be performed in the framework of a corresponding fuzzy space. Obscure-quantum computation may be considered as a set of rules (defining obscure-quantum gates) for managing quantum and membership amplitudes independently in different spaces. In this framework, we obtain not only the probabilities of final states but also their membership functions, i.e., how much trust we should assign to these probabilities. Our approach considerably extends the theory of quantum computing by adding the logic part directly to the computation process. Future challenges could lie in the direction of development of the corresponding logic hardware in parallel with the quantum devices.

References

Acampora G, Schiattarella R and Vitiello A 2023 On the implementation of fuzzy inference engines on quantum computers *IEEE Trans. Fuzzy Syst.* **31** 1419–33

Alexandru A, Bedaque P F, Carosso A, Cervia M J, Murairi E M and Sheng A 2023 *Fuzzy Gauge Theory for Quantum Computers* preprint Univ. Maryland College Park 16 p (arxiv:hep-lat/2308.05253)

Ávila A, Schmalfuss M, Reiser R and Kreinovich V 2015 Fuzzy Xor classes from quantum computing *Artificial Intelligence and Soft Computing* (Cham: Springer) 305–17

Barenco A, Bennett C H, Cleve R, DiVincenzo D P, Margolus N, Shor P, Sleator T, Smolin J A and Weinfurter H 1995 Elementary gates for quantum computation *Phys. Rev.* **52** 3457–67

Bartková R, Riečan B and Tirpáková A 2017 *Probability Theory for Fuzzy Quantum Spaces with Statistical Applications* (Sharjah: Bentham Science Publishers)

Battilotti G and Zizzi P 2004 *Logical Interpretation of a Reversible Measurement in Quantum Computing* preprint Univ. Padova Padova 25 p (quant-ph/0408068)

Benjamin S C 2001 Quantum computing without local control of qubit-qubit interactions *Phys. Rev. Lett.* **88** 017904

Bertini C and Leporini R 2017 A fuzzy approach to quantum logical computation *Fuzzy Sets Syst.* **317** 44–60

Bharill N, Patel O P and Tiwari A 2015 An enhanced quantum-inspired evolutionary fuzzy clustering *2015 IEEE Symposium Series on Computational Intelligence* (Cape Town, South Africa: IEEE) pp 772–9

Bharill N, Patel O P, Tiwari A, Mu L, Li D-L, Mohanty M, Kaiwartya O and Prasad M 2019 A generalized enhanced quantum fuzzy approach for efficient data clustering *IEEE Access* **7** 50347–61

Blutner R, Pothos E M and Bruza P 2013 A quantum probability perspective on borderline vagueness *Top. Cogn. Sci.* **5** 711–36

Bolotin A 2018 Truth values of quantum phenomena *Int. J. Theor. Phys.* **57** 2124–32

Boualem A, Runz C, Ayaida M and Akdag H 2023 A fuzzy/possibility approach for area coverage in wireless sensor networks *Soft Comput.* **27** 9367–82

Brylinski J L and Brylinski R 1994 universal quantum gates *Mathematics of Quantum Computation* (Boca Raton: Chapman & Hall/CRC Press) pp 124–34

Buckley J J 1989 Fuzzy complex numbers *Fuzzy Sets Syst.* **33** 333–45

Bugajski S 1998a Fuzzy stochastic processes *Open Syst. Inf. Dyn.* **5** 169–85

Bugajski S 1998b Net entropies of fuzzy stochastic processes *Open Syst. Inf. Dyn.* **5** 187–200

Bukhori I and Silitonga A 2014 A new approach of quantum-inspired genetic algorithm for self-generation of fuzzy logic controller *2014 International Conference on Intelligent Autonomous Agents, Networks and Systems* (Bandung, Indonesia: IEEE) pp 30–5

Busch P and Jaeger G 2010 Unsharp quantum reality *Found. Phys.* **40** 1341–67

Bělohlávek R 2002 *Fuzzy Relational Systems: Foundations and Principles* (New York: Springer)

Cattaneo G, Leporati A and Leporini R 2008 Quantum conservative many-valued computing *Fuzzy Sets Syst.* **159** 1001–30

Clark T and Hicks M G 2020 Models of necessity *Beilstein J. Org. Chem.* **16** 1649–61

Cunha R A F, Sharma N, Toffano Z and Dubois F 2019 Fuzzy logic behavior of quantum-controlled Braitenberg vehicle agents ed B Coecke and A Lambert-Mogiliansky *International Symposium on Quantum Interaction (QI 2018)* ed B Coecke and A Lambert-Mogiliansky (Berlin: Springer) pp 111–22

Dalla Chiara M, Giuntini R and Greechie R 2004 *Reasoning in Quantum Theory. Sharp and Unsharp Quantum Logics* vol 22 (Dordrecht: Kluwer)

Dalla Chiara M L and Giuntini R 1996 Fuzzy quantum logics *Mathware Soft Comput.* **3** 83–91

Dalla Chiara M L and Giuntini R 1999 Łukasiewicz' theory of truth, from the quantum logical point of view *Alfred Tarski and the Vienna circle* (Dordrecht: Kluwer) pp 127–34

Dalla Chiara M L, Giuntini R and Leporini R 2003 Quantum computational logics: a survey *Trends in Logic Trends Log. Stud. Log. Libr.* vol 21 (Dordrecht: Kluwer) pp 229–71

Dalla Chiara M L, Giuntini R, Sergioli G and Leporini R 2018 A many-valued approach to quantum computational logics *Fuzzy Sets Syst.* **335** 94–111

Deng J and Deng Y 2021 QZNs: Quantum Z-numbers *J. Latex Class Files* **14** 1–13

Deutsch D 1985 Quantum theory, the Church-Turing principle and the universal quantum computer *Proc. Roy. Soc. London* A400 96–117

Domenech G and Freytes H 2006 Fuzzy propositional logic associated with quantum computational gates *Int. J. Theor. Phys.* **45** 237–70

Dompere K K 2019 *A general theory of entropy. Fuzzy rational foundations of information-knowledge certainty vol 384 of Stud. Fuzziness Soft Comput.* (Cham: Springer)

Dubois D, Nguyen H T and Prade H 2000 Possibility theory, probability and fuzzy sets, misunderstandings, bridges and gaps *Fundamentals of Fuzzy Sets* ed D Dubois and H Prade (Boston: Springer) pp 343–438

Dubois E D and Prade H 2000 *Fundamentals of fuzzy sets vol 7 of The Handbooks of Fuzzy Sets Series* vol 7 (Boston: Kluwer)

Duplij S and Vogl R 2021 Obscure qubits and membership amplitudes *Topics on Quantum Information Science* ed S Curilef and A R Plastino (London: IntechOpen) p 20

Dvurečenskij A and Pulmannová S 2000 *New Trends in Quantum Structures* vol 516 (Dordrecht: Kluwer)

Dvurečenskij A and Chovanec F 1988 Fuzzy quantum spaces and compatibility *Int. J. Theor. Phys.* **27** 1069–82

El Naschie M S 2008 Fuzzy platonic spaces as a model for quantum physics *Math. Methods, Phys. Methods Simul. Sci. Tech.* **1** 91–101

Engesser K, Gabbay D M and Lehmann D (ed) 2009 *Handbook of Quantum Logic and Quantum Structures* (Amsterdam: Elsevier/North-Holland)

Freytes H, Giuntini R and Sergioli G 2019 Holistic type extension for classical logic via Toffoli quantum gate *Entropy* **21** 636

Freytes H, Giuntini R and Sergioli G 2010 Representing fuzzy structures in quantum computation with mixed states *2010 40th IEEE International Symposium on Multiple-Valued Logic* (Barcelona: IEEE) pp 162–6

Freytes H and Sergioli G 2014 Fuzzy approach for Toffoli gate in quantum computation with mixed states *Rep. Math. Phys.* **74** 159–80

Frič R 2021 *Tatra Mt. Math. Publ.* **78** 119–28

Frič R and Papčo M 2017 Probability: from classical to fuzzy *Fuzzy Sets Syst.* **326** 106–14

Garrido A 2012 Axiomatic of fuzzy complex numbers *Axioms* **1** 21–32

Genovese M and Traina P 2008 Review on qudits production and their application to quantum communication and studies on local realism *Adv. Sci. Lett.* **1** 153–60

Gentili P L 2009 Fuzzy logic in molecular computing *Int. J. Comp. Res.* **16** 283–92

Gentili P L 2011 Molecular processors: from qubits to fuzzy logic Chemphyschem: a European *J. Chem. Phys. Chem.* **12** 739–45

Gentili P L 2018 The fuzziness of the molecular world and its perspectives *Molecules* **23** 2074

Georgescu G 2004 Bosbach states on fuzzy structures *Soft Comput.* **8** 217–30

Goodman I R and Nguyen H T 2002 Fuzziness and randomness *Statistical Modeling Analysis and Management of Fuzzy Data* ed I R Goodman, H T Nguyen and D A Ralescu (Berlin: Springer)

Granik A 1994 Fuzziness in quantum mechanics *Adaptive Computing: Mathematics, Electronics and Optics: A Critical Review* ed S-S Chen and H J Caulfield (Orlando: SPIE) pp 228–49

Granik A and Caulfield H J 1996 Fuzziness in quantum mechanics *Phys. Essays* **9** 496–505

Gudder S 1986 Partial Hilbert spaces and amplitude functions *Ann. Ins. Henri Poincaré* **45** 311–26

Gudder S 1988 A theory of amplitudes *J. Math. Phys.* **29** 2020–35

Gudder S 1989 Fuzzy amplitude densities and stochastic quantum mechanics *Found. Phys.* **19** 293–317

Gudder S 2003 Quantum computational logic *Int. J. Theor. Phys.* **42** 39–47

Gudder S 2005 Foundations of probability and physics—3 *AIP Conf. Proc.* vol 750; A Khrennikov (Melville: Amer. Inst. Phys.)

Hannachi M S, Hatakeyama Y and Hirota K 2007a Emulating qubits with fuzzy logic *J. Adv. Comput. Intell. Intell. Informatics* **11** 242–9

Hannachi M S, Hatakeyama Y and Hirota K 2007b Emulating qubits with fuzzy logic *J. Adv. Comput. Intel. Inf.* **11** 242–9

Hannan M A, Ali J A, Hossain Lipu M S, Mohamed A, Ker P J, Indra Mahlia T M, Mansor M, Hussain A, Muttaqi K M and Dong Z Y 2020 Role of optimization algorithms based fuzzy controller in achieving induction motor performance enhancement *Nat. Commun.* **11** 3792

Hirota K and Ozawa K 1989 The concept of fuzzy flip-flop *IEEE Trans. Syst. Man Cybern.* **19** 980–97

Ho J K W and Hoorn J F 2022 Quantum affective processes for multidimensional decision-making *Sci. Rep.* **12** 20468

Ishikawa S 1997 A quantum mechanical approach to a fuzzy theory *Fuzzy Sets Syst.* **90** 277–306

Ishikawa S 1998 Fuzzy logic in measurements *Fuzzy Sets Syst.* **100** 291–300

Juels A and Wattenberg M 1999 A fuzzy commitment scheme *Proceedings of the 6th ACM conference on Computer and communications security* (New York: Association for Computing Machinery) pp 28–36

Kåhre J 2002 The Mathematical Theory of Information *Kluwer Int. Ser. Eng. Comput. Sci.* (Boston: Kluwer)

Kashmadze G 2017 Fuzzyfication of the Bloch ball *GESJ Comp. Sci. Telecomm.* **2017** 30–6

Kaye P, Laflamme R and Mosca M 2007 *An introduction to quantum computing* (Oxford: Oxford Univ. Press)

Kóczy L T and Hirota K 1990 Digital circuits based on algebraic fuzzy operations *Progress in fuzzy sets and systems (Vienna, 1989)* vol 5 ed W H Janko, M Roubens and H-J Zimmermann (Dordrecht: Kluwer) pp 100–14

Kosko B 1997 *Fuzzy Engineering* (Upper Saddle River: Prentice Hall, Inc)

Kreinovich V, Kohout L and Kim E 2011 Square root of 'not': a major difference between fuzzy and quantum logics *Int. J. General Systems* **40** 111–27

Kreinovich V and Kosheleva O 2021 Zadeh's vision, modern physics and the future of computing *TWMS J. Pure Appl. Math.* **12** 41–53

Leporini R, Bertini C and Fabiani F C 2020 Fuzzy representation of finite-valued quantum gates *Soft Comput.* **24** 10305–13

Li S, Li P and Yuan L 2007 Quantum genetic algorithm with application in fuzzy controller parameter optimization *Syst. Eng. Electronics* **29** 1134–8

Li Z 2013 On a new interpretation of fuzzy C mean using quantum theory *J. Algorithms Comput. Technol.* **7** 197–208

Litvintseva L V, Ul'yanov I S, Ul'yanov S V and Ul'yanov S S 2007 Quantum fuzzy inference for knowledge base design in robust intelligent controllers *J. Comput. Sys. Sc. Int.* **46** 908–61

Liu W and Li S 2011 On standard models of fuzzy region connection calculus *Int. J. Approximate Reasoning* **52** 1337–54

Liu Y and Luo S 2021 Quantifying unsharpness of measurements via uncertainty *Phys. Rev.* A **104** 10

Lloyd S 1995 Almost any quantum logic gate is universal *Phys. Rev. Lett.* **75** 346–9

Lu J, Han J, Hu Y and Zhang G 2016 Multilevel decision-making: A survey *Inform. Sci.* **346/347** 463–87

Mangaraj B K 2016 Relative effectiveness analysis under fuzziness *Procedia Comput. Sci.* **102** 231–8

Mannucci M A 2006 *Quantum Fuzzy sets: Blending Fuzzy Set Theory and Quantum Computation* preprint G. Mason Univ. Fairfax 12 p (arXiv:cs/0604064)

Maron A, Visintin L, Reiser R H S and Abeijon A M 2013 Fuzzy computing from quantum computing–case study in Reichenbach implication class *Mathware and Soft Comp. Mag.* **20** 86–93

Martín-Vide C, Mitrana V and Păun G (ed) 2004 *Formal languages and applications vol 148 of Stud. Fuzziness Soft Comput.* (Berlin: Springer)

Martínez-Garaot S, Pons M and Sokolovski D 2020 From quantum probabilities to quantum amplitudes *Entropy* **22** 15

Martsenyuk M A 2007 Matrix representation of fuzzy logic *Fuzzy Syst. Soft Comp.* **2** 7–35

Meyer C, Kandel A and Rundus D 1993 The triad of fuzzy theory *ACM SIGAPP Appl. Comp. Rev.* **1** 12–5

Mitra A 2022 Quantifying unsharpness of observables in an outcome-independent way *Int. J. Theor. Phys.* **61** 27

Mohrhoff U 2009 Objective probability and quantum fuzziness *Found. Phys.* **39** 137–55

Nguyen H T 2021 An enjoyable research journey on uncertainty *Statistical and Fuzzy Approaches to Data Processing, with Applications to Econometrics and Other Areas* (Cham: Springer)

Nguyen H T, Walker C L and Walker E A 2018 *A first course in fuzzy logic* 4th edn (Boca Raton: CRC Press)

Nielsen M A and Chuang I L 2000 *Quantum Computation and Quantum Information* (Cambridge: Cambridge University Press)

Patel O P, Bharill N, Tiwari A and Prasad M 2021 A novel quantum-inspired fuzzy based neural network for data classification *IEEE Trans. Emerg. Topics Comput.* **9** 1031–44

Pineda C, Davalos D, Viviescas C and Rosado A 2020 Fuzzy measurements and coarse graining in quantum many-body systems *Phys. Rev.* A **104** 042218

Pratihar J, Dey A, Khan A, Banerjee P and Pal R K 2023 Computing with words for solving the fuzzy transportation problem *Soft Comput.* **1** 1–14

Prugovečki E 1977 Information-theoretical aspects of quantum measurement *Int. J. Theor. Phys.* **16** 321–31

Pykacz J 2000 Quantum logic as a basis for computations *Int. J. Theor. Phys.* **39** 839–50

Pykacz J 2006 Solution' of the EPR paradox: negative, or rather fuzzy probabilities? *Found. Phys.* **36** 437–42

Pykacz J 2015 *Quantum Physics, Fuzzy Sets and Logic* (Cham: Springer)

Qiu W 2020 There also can be fuzziness in quantum states itself-breaking through the framework and the principle of quantum mechanics *J. Mod. Phys.* **11** 952–66

Qiu W 2022 Fuzziness in quantum states—breaking through the framework and the principle of quantum mechanics *New Trends in Physical Science Research* vol 6 (London: Book Publisher International) pp 12–28

Ramot D, Milo R, Friedman M and Kandel A 2002 *IEEE Trans. Fuzzy Syst.* **10** 171–86

Regan K, Chakrabarti A and Guan C 2018 Algebraic and logical emulations of quantum circuits *Transactions on Computational Science XXXI. Special issue on signal processing and security in distributed systems* (Berlin: Springer) pp 41–76

Reiser R, Lemke A, Avila A, Vieira J, Pilla M and Du Bois A 2016 Interpretations on quantum fuzzy computing: intuitionistic fuzzy operations × quantum operators *Electron. Notes Theor. Comput. Sci.* **324** 135–50

Rigatos G G 2007 Quantum wave-packets in fuzzy automata and neural associative memories *Internat. J. Modern Phys.* C **18** 1551–69

Rocha A F, Massad E and Pereira A 2005 The Brain: Fuzzy Arithmetic to Quantum Computing *Stud. Fuzziness Soft Comput.* vol 165 (Berlin: Springer)

Roy S, Kundu M K and Granlund G H 1996 Uncertainty relations and time-frequency distributions for unsharp observables *Inform. Sci.* **89** 193–209

Ruspini E H, Bonissone P and Pedrycz W 1998 *Handbook of Fuzzy Computation* (Boca Raton: CRC Press)

Rybalov A, Kagan E, Rapoport A and Ben-Gal I 2014 Fuzzy implementation of qubits operators *J. Comput. Sci. J. Comput. Sci.* **07** 163–8

Sander W 1994 Uncertainty and fuzziness *Fuzzy-Systems in Computer Science* (Wiesbaden: Vieweg-Teubner Verlag) pp 265–73

Schmitt I, Nürnberger A and Lehrack S 2009 On the relation between fuzzy and quantum logic *Views on Fuzzy Sets and Systems from Different Perspectives Philosophy and Logic, Criticisms and Applications* (Berlin: Springer) pp 417–38

Seising R 2006 Can fuzzy sets be useful in the (re) interpretation of uncertainty in quantum mechanics? *2006 Annual Meeting of the North American Fuzzy Information Processing Society* (Montreal: IEEE) 414–9

Sergioli G and Freytes H 2016 Fuzzy approach for CNOT gate in quantum computation with mixed states *Probing the Meaning of Quantum Mechanics. Superpositions, Dynamics, Semantics and Identity* (Hackensack: World Scientific) pp 244–58

Sergioli G, Giuntini R and Paoli F 2011 Irreversibility in quantum computational logics *Appl. Math. Inform. Sci.* **5** 171–91

Smith N J J 2008 *Vagueness and Degrees of Truth* (Oxford: Oxford University Press)

Song Q and Liu L 2012 Fuzzy quantum genetic algorithm for solving global optimization problem *J. Converg. Inf. Tech.* **7** 460–7

Syropoulos A 2017 On vague computers *Emergent Computation* (Cham: Springer) pp 393–402

Toffano Z and Dubois F 2017 Eigenlogic: Interpretable quantum observables with applications to fuzzy behavior of vehicular robots *Science with and for Society: Contributions of Cybernetics and Systems* (Rome) pp 1–5

Ulyanov S V 2020 Quantum fuzzy inference based on quantum genetic algorithm: Quantum simulator in intelligent robotics *Advances in Intelligent Systems and Computing* (Berlin: Springer) pp 78–85

Virant J 2000 *Design Considerations of Time in Fuzzy Systems* vol 35 (Dordrecht: Kluwer)

Wang Y, Hu Z, Sanders B C and Kais S 2020 *Qudits and High-Dimensional Quantum Computing* preprint Purdue University West Lafayette p. (arxiv:2008.00959)

Williams C P and Clearwater S H 1998 *Explorations in Quantum Computing* (New York-Berlin-Heidelberg: Springer)

Yakubova N and Abdurasulova G 2023 Investigation of fuzzy controllers in intelligent control systems based on quantum computing *Universum: Tech. Sci.* **108** 22–5

Yan P, Jiang H and Yu N 2022 On incorrectness logic for quantum programs *Proc. ACM Program. Lang.* vol 6 (New York: Association for Computing Machinery (ACM)) pp 1–28

Zadeh L A 1996a Fuzzy algorithms *Advances in Fuzzy Systems — Applications and Theory* (Singapore: World Scientific) pp 60–8

Zadeh L A 1996b Fuzzy sets *Advances in Fuzzy Systems—Applications and Theory* (Singapore: World Scientific) 394–432

Zadeh L A 1998 Soft computing, fuzzy logic and recognition technology *1998 IEEE Int. Conf. Fuzzy Systems Proc. IEEE World Congress on Computational Intelligence (Cat. No. 98CH36228)* vol 2 (Anchorage: IEEE) pp 1678–9

Zhou B, Xu G and Li S 2015 The quintuple implication principle of fuzzy reasoning *Inform. Sci.* **297** 202–15

Zimmermann H-J 2011 *Fuzzy Set Theory and Its Applications* 2nd eda (New York: Springer)

Zizzi P 2005 Qubits and quantum spaces *Int. J. Quant. Inf* **3** 287–91

Zizzi P and Pessa E 2014 From SU(2) gauge theory to qubits on the fuzzy sphere *Int. J. Theor. Phys.* **53** 257–70

Chapter 2

Higher braid quantum gates

One of the main problems in the physical realization of quantum computers is the presence of errors, which implies that it is desirable for quantum computations be provided with error correction or that ways be found to make the states more stable, which leads to the concept of topological quantum computation (for reviews, see, e.g., Freedman *et al* 2003, Nayak *et al* 2008, Rowell and Wang 2018, and references therein). In the Turaev approach (Turaev 1988), link invariants can be obtained from the solutions of the constant Yang–Baxter equation (the braid equation). It was realized that the topological entanglement of knots and links is deeply connected with quantum entanglement (Aravind 1997, Kauffman and Lomonaco 2002). Indeed, if the solutions to the constant Yang–Baxter equation (Lambe and Radford 1997) or Yang–Baxter operators/maps (Bukhshtaber 1998, Veselov 2003) are interpreted as a special class of quantum gate, namely braiding quantum gates (Kauffman and Lomonaco 2004, Melnikov *et al* 2018), then the inclusion of non-entangling gates does not change the relevant topological invariants (Alagic *et al* 2016, Kauffman and Mehrotra 2019). For further properties and applications of braiding quantum gates, see Melnikov *et al* (2019), Ballard and Wu (2011b), Kolganov and Morozov (2020), and Kolganov *et al* (2021).

Here we obtain and study (Duplij and Vogl 2021) the solutions to the higher arity (polyadic) braid equations introduced in Duplij (2021b, 2021a) as a polyadic generalization of the constant Yang–Baxter equation, which is considerably different from the generalized Yang–Baxter equation of Rowell *et al* (2010), Kitaev and Wang (2012), Vasquez *et al* (2016), and Padmanabhan *et al* (2020b). We introduce special classes of matrices (star and circle types), to which most of the solutions belong, and find that the so-called magic matrices (Khaneja and Glaser 2001, Kraus and Cirac 2001, Ballard and Wu 2011b) belong to the star class. We investigate their general non-trivial group properties and polyadic generalizations. We then consider the invertible and noninvertible matrix solutions to the higher braid equations as the corresponding higher braiding gates acting on multi-qubit states. It is important for

doi:10.1088/978-0-7503-5281-9ch2

multi-qubit entanglement can speed up quantum key distribution (Epping *et al* 2017) and accelerate various algorithms (Vartiainen *et al* 2004). As an example, we have made detailed computations for the ternary braiding gates as solutions to the ternary braid equations (Duplij 2021b, 2021a). A particular solution to the n-ary braid equation is also presented. It is shown that for each multi-qubit state, there exist higher braiding gates that are not entangling and the concrete relations for that are obtained, which can allow us to build non-entangling networks.

2.1 Yang–Baxter operators

Recall here (Kauffman and Lomonaco 2002, 2004) the standard construction of the special kind of gates we will consider, the braiding gates, in terms of solutions to the *constant Yang–Baxter equation* (Lambe and Radford 1997), also called the *algebraic* Yang–Baxter equation (Dye 2003), or the (binary) *braid equation* (Duplij 2021b).

2.1.1 Yang–Baxter maps and braid group

First we consider a general abstract construction of the (binary) braid equation. Let V be a vector space over a field $\mathbb{K}$ and the mapping $C_{V^2}\colon V\otimes V\to V\otimes V$, where $\otimes=\otimes_{\mathbb{K}}$ is the tensor product over $\mathbb{K}$. A linear operator (*braid operator*) C_{V^2} is called a *Yang–Baxter operator* (denoted by R in Kauffman and Lomonaco 2004 and by B in Lambe and Radford 1997) or *Yang–Baxter map* (Veselov 2003) (denoted by F in Bukhshtaber 1998) if it satisfies the *braid equation* (Drinfeld 1989, 1992, Kassel 1995).

$$(C_{V^2}\otimes \mathrm{id}_V)\circ(\mathrm{id}_V\otimes C_{V^2})\circ(C_{V^2}\otimes \mathrm{id}_V)=(\mathrm{id}_V\otimes C_{V^2})\circ(C_{V^2}\otimes \mathrm{id}_V)\circ(\mathrm{id}_V\otimes C_{V^2}) \tag{2.1}$$

where $\mathrm{id}_V\colon V\to V$ is the identity operator in V. The connection of C_{V^2} with the R-matrix R is given by $C_{V^2}=\tau\circ R$, where τ is the flip operation (Drinfeld 1989, Bukhshtaber 1998, Lambe and Radford 1997).

Let us introduce the operators $A_{1,2}\colon V\otimes V\otimes V\to V\otimes V\otimes V$ by

$$A_1=C_{V^2}\otimes \mathrm{id}_V,\quad A_2=\mathrm{id}_V\otimes C_{V^2}, \tag{2.2}$$

It follows from (2.1) that

$$A_1\circ A_2\circ A_1=A_2\circ A_1\circ A_2. \tag{2.3}$$

If C_{V^2} is invertible, then $C_{V^2}^{-1}$ is also the Yang–Baxter map with A_1^{-1} and A_2^{-1}. Therefore, the operators A_i represent the braid group $\mathscr{B}_3=\{e,\sigma_1,\sigma_2|\sigma_1\sigma_2\sigma_1=\sigma_2\sigma_1\sigma_2\}$ by the mapping π_3 as

$$\mathscr{B}_3\xrightarrow{\pi_3}\mathrm{End}(V\otimes V\otimes V),\quad \sigma_1\overset{\pi_3}{\mapsto}A_1,\quad \sigma_2\overset{\pi_3}{\mapsto}A_2,\quad e\overset{\pi_3}{\mapsto}\mathrm{id}_V. \tag{2.4}$$

The representation π_m of the braid group with m strands

$$\mathscr{B}_m=\left\{e,\sigma_1,\ldots,\sigma_{m-1}\,\middle|\,\begin{array}{c}\sigma_i\sigma_{i+1}\sigma_i=\sigma_{i+1}\sigma_i\sigma_{i+1},\ i=1,\ldots,m-1,\\ \sigma_i\sigma_j=\sigma_j\sigma_i,\quad |\,i-j\,|\geqslant 2,\end{array}\right\} \tag{2.5}$$

can be obtained using operators $A_i(m)$: V $^{\otimes m}\rightarrow$ V $^{\otimes m}$ analogous to (2.2)

$$A_i(m) = \overbrace{\mathrm{id}_V \otimes \cdots \otimes \mathrm{id}_V}^{i-1} \otimes C_{V^2} \otimes \overbrace{\mathrm{id}_V \otimes \cdots \mathrm{id}_V}^{m-i-1}, \tag{2.6}$$
$$A_0(m) = (\mathrm{id}_V)^{\otimes m}, \quad i = 1, \cdots, m-1,$$

by the mapping π_m: $\mathscr{B}_m \rightarrow$ End V $^{\otimes m}$ in the following way

$$\pi_m(\sigma_i) = A_i(m), \qquad \pi_m(e) = A_0(m). \tag{2.7}$$

In this notation (2.2) is $A_i = A_i(2)$, $i = 1, 2$, and therefore (2.3) represents $\mathscr{B}_3$ by (2.4).

2.1.2 Constant matrix solutions to the Yang–Baxter equation

Consider next a concrete version of the vector space V that is used in the quantum computation, a d-dimensional Euclidean vector space V_d over complex numbers $\mathbb{C}$ with a basis $\{e_i\}$, $i = 1, \dots, d$. A linear operator $V_d \rightarrow V_d$ is given by a complex $d \times d$ matrix, the identity operator id_V becomes the identity $d \times d$ matrix I_d, and the Yang–Baxter map C_{V^2} is a $d^2 \times d^2$ matrix C_{d^2} (denoted by R in Dye 2003) satisfying the matrix algebraic Yang–Baxter equation

$$(C_{d^2} \otimes I_d)(I_d \otimes C_{d^2})(C_{d^2} \otimes I_d) = (I_d \otimes C_{d^2})(C_{d^2} \otimes I_d)(I_d \otimes C_{d^2}), \tag{2.8}$$

being an equality between two matrices of size $d^3 \times d^3$. We use the unified notations, which can be straightforwardly generalized for higher braid operators. In components

$$C_{d^2} \circ (e_{i_1} \otimes e_{i_2}) = \sum_{j_1', j_2'=1}^{d} c_{i_1 i_2}^{\ j_1' j_2'} \cdot e_{j_1'} \otimes e_{j_2'}, \tag{2.9}$$

the Yang–Baxter equation (2.8) has the shape (where summing is by primed indices)

$$\sum_{j_1', j_2', j_3'=1}^{d} c_{i_1 i_2}^{\ j_1' j_2'} \cdot c_{j_2' i_3}^{\ j_3' k_3} \cdot c_{j_1' j_3'}^{\ k_1 k_2} = \sum_{l_1', l_2', l_3'=1}^{d} c_{i_2 i_3}^{\ l_2' l_3'} \cdot c_{i_1 l_2'}^{\ k_1 l_1'} \cdot c_{l_1' l_3'}^{\ k_2 k_3} \equiv q_{i_1 i_2 i_3}^{\ k_1 k_2 k_3}. \tag{2.10}$$

The system (2.10) is highly overdetermined because the matrix C_{d^2} contains d^4 unknown entries, while there are d^6 cubic polynomial equations for them. So for $d = 2$ we have 64 equations for 16 unknowns, while for $d = 3$ there are 729 equations for the 81 unknown entries of C_{d^2}. The unitarity of C_{d^2} imposes a further d^2 quadratic equations, and so for $d = 2$ we have in total 68 equations for 16 unknowns. This makes the direct discovery of solutions for the matrix Yang–Baxter equation (2.10) very cumbersome. Nevertheless, using a conjugation classes method, the unitary solutions and their classification for $d = 2$ were presented in Dye (2003).

In the standard matrix form, (2.9) can be presented by introducing the four-dimensional vector space $\tilde{V}_4 = V \otimes V$ with the natural basis $\tilde{e}_{\tilde{k}} = \{e_1 \otimes e_1, e_1 \otimes e_2, e_2 \otimes e_1, e_2 \otimes e_2\}$, where $\tilde{k} = 1, \dots, 8$ is a cumulative index.

The linear operator $\tilde{C}_4$: $\tilde{V}_4 \to \tilde{V}_4$ corresponding to (2.9) is given by 4×4 matrix $\tilde{c}_{\tilde{\imath}\tilde{\jmath}}$ as $\tilde{C}_4 \circ \tilde{e}_{\tilde{\imath}} = \sum_{\tilde{\jmath}=1}^{4} \tilde{c}_{\tilde{\imath}\tilde{\jmath}} \cdot \tilde{e}_{\tilde{\jmath}}$. The operators (2.2) become two 8×8 matrices $\tilde{A}_{1,2}$ as

$$\tilde{A}_1 = \tilde{c} \otimes_{\mathrm{K}} I_2, \quad \tilde{A}_2 = I_2 \otimes_{\mathrm{K}} \tilde{c}, \tag{2.11}$$

where $\otimes_{\mathrm{K}}$ is the Kronecker product of matrices and I_2 is the 2×2 identity matrix. In this notation, which is universal and also used for higher braid equations, the operator binary braid equations (2.117) become a single matrix equation

$$\tilde{A}_1 \tilde{A}_2 \tilde{A}_1 = \tilde{A}_2 \tilde{A}_1 \tilde{A}_2, \tag{2.12}$$

which we call the *matrix binary braid equation*, and also the constant Yang–Baxter equation (Dye 2003). In component form, (2.12) is a highly overdetermined system of 64 cubic equations for 16 unknowns, the entries of $\tilde{c}$.

The matrix equation (2.12) has the following gauge invariance, which allows a classification of Yang–Baxter maps (Hietarinta 1993). Introduce an invertible operator Q: $V \to V$ in the two-dimensional vector space $V \equiv V_{d=2}$. In the basis $\{e_1, e_2\}$, its 2×2 matrix q is given by $Q \circ e_i = \sum_{j=1}^{2} q_{ij} \cdot e_j$. In the natural four-dimensional basis $\tilde{e}_{\tilde{k}}$ the tensor product of operators $Q \otimes Q$ is presented by the Kronecker product of matrices $\tilde{q}_4 = q \otimes_{\mathrm{K}} q$. If the 4×4 matrix $\tilde{c}$ is a fixed solution to the Yang–Baxter equation (2.12), then the family of solutions $\tilde{c}(q)$ corresponding to the invertible 2×2 matrix q is the conjugation of $\tilde{c}$ by $\tilde{q}_4$ such that

$$\tilde{c}(q) = \tilde{q}_4 \tilde{c} \tilde{q}_4^{-1} = (q \otimes_{\mathrm{K}} q) \tilde{c} (q^{-1} \otimes_{\mathrm{K}} q^{-1}), \tag{2.13}$$

which follows from conjugating (2.12) by $q \otimes_{\mathrm{K}} q$ and using (2.11). If we include the obvious invariance of (2.12) with respect to an overall factor $t \in \mathbb{C}$, then the general family of solutions becomes (cf the Yang–Baxter equation Hietarinta 1993)

$$\tilde{c}(q, t) = t \tilde{q}_4 \tilde{c} \tilde{q}_4^{-1} = t (q \otimes_{\mathrm{K}} q) \tilde{c} (q^{-1} \otimes_{\mathrm{K}} q^{-1}). \tag{2.14}$$

It follows from (2.13) that the matrix $q \in \mathrm{GL}(2, \mathbb{C})$ is defined up to a complex nonzero factor. In this case we can put

$$q = \begin{pmatrix} a & 1 \\ c & d \end{pmatrix}, \tag{2.15}$$

and the manifest form of $\tilde{q}_4$ is

$$\tilde{q}_4 = \begin{pmatrix} a^2 & a & a & 1 \\ ac & ad & c & d \\ ac & c & ad & d \\ c^2 & cd & cd & d^2 \end{pmatrix}. \tag{2.16}$$

The matrix $\tilde{q}_4^{\star} \tilde{q}_4$ (where $\star$ represents Hermitian conjugation) is diagonal (this case is important in a further classification similar to the binary one Dye 2003), when the condition

$$c = -a/d^* \tag{2.17}$$

holds, and so the matrix q takes the special form (depending on two complex parameters)

$$q = \begin{pmatrix} a & 1 \\ -a/d^* & d \end{pmatrix}. \tag{2.18}$$

We call two solutions $\tilde{c}_1$ and $\tilde{c}_2$ of the constant Yang–Baxter equation (2.12) *q-conjugated*, if

$$\tilde{c}_1\tilde{q}_4 = \tilde{q}_4\tilde{c}_2, \tag{2.19}$$

and we will not distinguish between them. The q-conjugation in the form (2.19) does not require the invertibility of the matrix q, and therefore the solutions of different ranks (or invertible and not invertible) can be q-conjugated (for the invertible case, see Hietarinta 1993, Alagic *et al* 2014, Padmanabhan *et al* 2021).

The matrix equation (2.12) does not imply the invertibility of solutions, i.e., matrices $\tilde{c}$ being of full rank (in the binary Yang–Baxter case of rank 4 and $d = 2$). Therefore, below we introduce in a unified way invertible and noninvertible solutions to the matrix Yang–Baxter equation (2.10) for any rank of the corresponding matrices.

2.1.3 Partial identity and unitarity

To be as close as possible to the invertible case, we introduce noninvertible analogs of identity and unitarity. Let M be a diagonal $n \times n$ matrix of rank $r \leqslant n$, and therefore with $n - r$ zeros on the diagonal. If the other diagonal elements are units, such a diagonal M can be reduced by row operations to a block matrix, being a direct sum of the identity matrix $I_{r\times r}$ and the zero matrix $Z_{(n-r)\times(n-r)}$.

Definition 2.1. We call such a diagonal matrix a *block r-partial identity* $I_n^{(\mathrm{block})}(r) = \mathrm{diag}\left\{\overbrace{1, \ldots 1}^{r}, \overbrace{0, \ldots, 0}^{n-r}\right\}$, and without the block reduction a *shuffler-partial identity* $I_n^{(\mathrm{shuffle})}(r)$ (these are connected by conjugation). We will use the term partial identity and $I_n(r)$ to denote any matrix of this form.

Obviously, with the full rank $r = n$ we have $I_n(n) \equiv I_n$, where I_n is the identity $n \times n$ matrix. As with the invertible case and identities, the partial identities (of the corresponding form) are *trivial solutions* of the Yang–Baxter equation.

Innovation 2.2. *If a matrix $M = M(r)$ of size $n \times n$ and rank r satisfies the following r-partial unitarity condition*

$$M(r)^\star M(r) = I_n^{(1)}(r), \tag{2.20}$$

$$M(r)M(r)^{\star} = I_n^{(2)}(r), \tag{2.21}$$

where $M(r)^{\star}$ *is the conjugate-transposed matrix and* $I_n^{(1)}(r)$, $I_n^{(2)}(r)$ *are partial identities (of any kind, they can be different), then* $M(r)$ *is called a r-partial unitary matrix.*

In the case, when $I_n^{(1)}(r) = I_n^{(2)}(r)$, the matrix $M(r)$ is called *normal*. If $M(r)^{\star} = M(r)$, then it is called *r-partial self-adjoint*. In the case of full rank $r = n$, the conditions (2.20)–(2.21) become ordinary unitarity, and $M(n)$ becomes an unitary (and normal) matrix, while a r-partial self-adjoint matrix becomes a self-adjoint matrix or Hermitian matrix.

As an example, we consider a 4×4 matrix of rank 3

$$M(3) = \begin{pmatrix} 0 & 0 & 0 & 0 \\ 0 & e^{i\beta} & 0 & 0 \\ 0 & 0 & 0 & e^{i\gamma} \\ e^{i\alpha} & 0 & 0 & 0 \end{pmatrix}, \quad \alpha, \beta, \gamma \in \mathbb{R}, \tag{2.22}$$

which satisfies the 3-partial unitarity conditions (2.20)–(2.21) with two different 3-partial identities on the rhs

$$M(3)^{\star}M(3) = \begin{pmatrix} 1 & 0 & 0 & 0 \\ 0 & 1 & 0 & 0 \\ 0 & 0 & 0 & 0 \\ 0 & 0 & 0 & 1 \end{pmatrix} = I_4^{(1)}(3) \neq I_4^{(2)}(3) = \begin{pmatrix} 0 & 0 & 0 & 0 \\ 0 & 1 & 0 & 0 \\ 0 & 0 & 1 & 0 \\ 0 & 0 & 0 & 1 \end{pmatrix} = M(3)M(3)^{\star}. \tag{2.23}$$

For a noninvertible matrix $M(r)$, one can define a *pseudoinverse* $M(r)^{+}$ (or the *Moore-Penrose inverse*) (Nashed 1976) by

$$M(r)M(r)^{+}M(r) = M(r), \quad M(r)^{+}M(r)M(r)^{+} = M(r)^{+}, \tag{2.24}$$

and $M(r)M(r)^{+}$, $M(r)^{+}M(r)$ are Hermitian. In the case of (2.22) the partial unitary matrix $M(3)$ coincides with its pseudoinverse

$$M(3)^{\star} = M(3)^{+}, \tag{2.25}$$

which is similar to the standard unitarity $M_{\text{inv}}^{\star} = M_{\text{inv}}^{-1}$ for an invertible matrix M_{inv}. It is important that (2.22) is a solution of the matrix Yang–Baxter equation (2.12)), and so is an example of a noninvertible Yang–Baxter map.

If only the first (second) of the conditions (2.20)–(2.21) holds, then we call such $M(r)$ a *left (right)r-partial unitary matrix*. An example of such a noninvertible Yang–Baxter map of rank 2 is the left 2-partial unitary matrix

$$M(2) = \frac{1}{\sqrt{2}} \begin{pmatrix} 0 & 0 & 0 & e^{i\alpha} \\ 0 & e^{i\beta} & 0 & 0 \\ 0 & e^{i\beta} & 0 & 0 \\ 0 & 0 & 0 & e^{i\beta} \end{pmatrix}, \quad \alpha, \beta \in \mathbb{R}, \tag{2.26}$$

which satisfies (2.20), but not (2.21), and so $M(2)$ is not normal

$$M(2)^{\star}M(2)=\begin{pmatrix}0&0&0&0\\0&1&0&0\\0&0&0&0\\0&0&0&1\end{pmatrix}\neq\begin{pmatrix}1&0&0&e^{i(\alpha-\beta)}\\0&1&1&0\\0&1&1&0\\e^{i(\beta-\alpha)}&0&0&1\end{pmatrix}=M(2)M(2)^{\star}. \tag{2.27}$$

Nevertheless, the property (2.25) still holds and $M(2)^{\star}=M(2)^{+}$.

2.1.4 Permutation and parameter-permutation 4-vertex Yang–Baxter maps

The system (2.12) with respect to all 16 variables is too cumbersome for direct solution. The classification of all solutions can only be accomplished in special cases, e.g., for matrices over finite fields (Hietarinta 1993) or for fewer than 16 vertices. Here we will start from 4-vertex permutation and parameter-permutation matrix solutions and investigate their group structure. It has been shown by Dye (2003), and Kauffman and Lomonaco (2004) that the special 8-vertex solutions to the Yang–Baxter equation are most important for further applications including braiding gates. We will therefore study the 8-vertex solutions in the most general way: over $\mathbb{C}$ and in various configurations, invertible and not invertible, and also consider their group structure.

First, we introduce the *permutation Yang–Baxter maps* that are presented by the permutation matrices (binary matrices with a single 1 in each row and column), i.e., 4-vertex solutions. In total, there are 64 permutation matrices of size 4×4, while only four of them have the full rank 4 and simultaneously satisfy the Yang–Baxter equation (2.12), as follows

$$\tilde{c}^{\,\mathrm{perm}}_{\mathrm{bisymm}}=\begin{pmatrix}1&0&0&0\\0&0&1&0\\0&1&0&0\\0&0&0&1\end{pmatrix},\begin{pmatrix}0&0&0&1\\0&1&0&0\\0&0&1&0\\1&0&0&0\end{pmatrix},\quad \begin{matrix}\operatorname{tr}\tilde{c}=2,\ \det\tilde{c}=-1,\\ \text{eigenvalues: }\{1\}^{[2]},\ \{-1\}^{[2]},\end{matrix} \tag{2.28}$$

$$\tilde{c}^{\,\mathrm{perm}}_{\mathrm{90symm}}=\begin{pmatrix}0&1&0&0\\0&0&0&1\\1&0&0&0\\0&0&1&0\end{pmatrix},\begin{pmatrix}0&0&1&0\\1&0&0&0\\0&0&0&1\\0&1&0&0\end{pmatrix},\quad \begin{matrix}\operatorname{tr}\tilde{c}=0,\ \det\tilde{c}=-1,\\ \text{eigenvalues: }1,\ i,\ -1,\ -i.\end{matrix} \tag{2.29}$$

Here and next we list the eigenvalues to understand which matrices are conjugated, and, after that, if and only if the conjugation matrix is of the form (2.16), then such solutions to the Yang–Baxter equation (2.12) coincide. The traces are important in the construction of corresponding link invariants (Turaev 1988) and local invariants (Balakrishnan and Sankaranarayanan 2010, Sudbery 2001), and the determinants are connected with the concurrence (Jaffali and Oeding 2020, Walter *et al* 2016). Note that the first matrix in (2.28) is the SWAP quantum gate (Nielsen and Chuang 2000).

To understand the symmetry properties of (2.28)–(2.29), we introduce the so-called *reverse matrix* $J\equiv J_n$ of size $n\times n$ by $(J_n)_{ij}=\delta_{i,n+1-i}$. For $n=4$ it is

$$J_4 = \begin{pmatrix} 0 & 0 & 0 & 1 \\ 0 & 0 & 1 & 0 \\ 0 & 1 & 0 & 0 \\ 1 & 0 & 0 & 0 \end{pmatrix}. \tag{2.30}$$

For any $n \times n$ matrix $M \equiv M_n$ the matrix JM is the matrix M reflected vertically, and the product MJ is M reflected horizontally. In addition to the standard *symmetric matrix* satisfying $M = M^{\mathrm{T}}$ (T is the transposition), one can introduce

$$M \text{is persymmetric, if } JM = (JM)^T, \tag{2.31}$$

$$M \text{is } 90° - \text{symmetric, if } M^{\mathrm{T}} = JM. \tag{2.32}$$

Thus, a persymmetric matrix is symmetric with respect to the minor diagonal, while a 90°-symmetric matrix is symmetric under 90°-rotations. A *bisymmetric matrix* is symmetric and persymmetric simultaneously. In this notation, the first family of the permutation solutions (2.28) are bisymmetric but not 90°-symmetric, while the second family of the solutions (2.29) are, oppositely, 90°-symmetric but not symmetric and not persymmetric (which explains their notation).

In the next step, we define the corresponding *parameter-permutation solutions* replacing the units in (2.28) with parameters. We found the following four 4-vertex solutions to the Yang–Baxter equation (2.12)) over $\mathbb{C}$

$$\tilde{c}^{\,\mathrm{perm,star}}_{\mathrm{rank}=4}(x, y, z, t) = \begin{pmatrix} x & 0 & 0 & 0 \\ 0 & 0 & y & 0 \\ 0 & z & 0 & 0 \\ 0 & 0 & 0 & t \end{pmatrix}, \begin{pmatrix} 0 & 0 & 0 & y \\ 0 & x & 0 & 0 \\ 0 & 0 & t & 0 \\ z & 0 & 0 & 0 \end{pmatrix},$$
$$\mathrm{tr}\, \tilde{c} = x + t,$$
$$\det \tilde{c} = -xyzt, \quad x, y, z, t \neq 0, \tag{2.33}$$
$$\text{eigenvalues: } x, t, \sqrt{yz}, -\sqrt{yz},$$

$$\tilde{c}^{\,\mathrm{perm,circ}}_{\mathrm{rank}=4}(x, y) = \begin{pmatrix} 0 & 0 & x & 0 \\ y & 0 & 0 & 0 \\ 0 & 0 & 0 & x \\ 0 & y & 0 & 0 \end{pmatrix}, \begin{pmatrix} 0 & x & 0 & 0 \\ 0 & 0 & 0 & x \\ y & 0 & 0 & 0 \\ 0 & 0 & y & 0 \end{pmatrix},$$
$$\mathrm{tr}\, \tilde{c} = 0,$$
$$\det \tilde{c} = -x^2 y^2, \quad x, y \neq 0, \tag{2.34}$$
$$\text{eigenvalues: } \sqrt{xy}, -\sqrt{xy}, i\sqrt{xy}, -i\sqrt{xy}.$$

Innovation 2.3. *The first pair of solutions (2.33) correspond to the bisymmetric permutation matrices (2.28), and we call them star-like solutions, while the second two solutions (2.34) correspond to the 90°-symmetric matrices (2.28), which are called circle-like solutions.*

The first (second) star-like solution in (2.33) with $y = z$ ($x = t$) becomes symmetric (persymmetric), while on the other hand with $x = t$ ($y = z$) it becomes persymmetric (symmetric). They become bisymmetric parameter-permutation solutions if all of the parameters are equal $x = y = z = t$. The circle-like solutions (2.34) are 90°-symmetric when $x = y$.

Using q-conjugation (2.14), one can next get families of solutions depending on the entries of q and the additional complex parameters in (2.15).

2.1.5 Group structure of 4-vertex and 8-vertex matrices

Let us analyze the group structure of 4-vertex matrices (2.33)–(2.34) with respect to matrix multiplication, i.e., which kinds of subgroups in GL(4, $\mathbb{C}$) they can form. For this we introduce four 4-vertex 4×4 matrices over $\mathbb{C}$: two star-like matrices

$$N_{\text{star1}} = \begin{pmatrix} x & 0 & 0 & 0 \\ 0 & 0 & y & 0 \\ 0 & z & 0 & 0 \\ 0 & 0 & 0 & t \end{pmatrix}, \qquad N_{\text{star2}} = \begin{pmatrix} 0 & 0 & 0 & y \\ 0 & x & 0 & 0 \\ 0 & 0 & t & 0 \\ z & 0 & 0 & 0 \end{pmatrix}, \quad \begin{array}{c} \operatorname{tr} N = x + t, \\ \det N = -xyzt, \quad x, y, z, t \neq 0, \\ \text{eigenvalues: } x, t, \sqrt{yz}, -\sqrt{yz}, \end{array} \tag{2.35}$$

and two circle-like matrices

$$N_{\text{circ1}} = \begin{pmatrix} 0 & 0 & x & 0 \\ y & 0 & 0 & 0 \\ 0 & 0 & 0 & z \\ 0 & t & 0 & 0 \end{pmatrix}, \qquad N_{\text{circ2}} = \begin{pmatrix} 0 & x & 0 & 0 \\ 0 & 0 & 0 & y \\ z & 0 & 0 & 0 \\ 0 & 0 & t & 0 \end{pmatrix}, \quad \begin{array}{c} \operatorname{tr} N = 0, \\ \det N = -xyzt, \quad x, y, z, t \neq 0, \\ \text{eigenvalues: } \sqrt[4]{xyzt}, -\sqrt[4]{xyzt}, i\sqrt[4]{xyzt}, -\sqrt[4]{xyzt}, \end{array} . \tag{2.36}$$

Denoting the corresponding sets by $\mathsf{N}_{\text{star1, 2}} = \{N_{\text{star1, 2}}\}$ and $\mathsf{N}_{\text{circ1, 2}} = \{N_{\text{circ1, 2}}\}$, these do not intersect and are closed with respect to the following multiplications

$$\mathsf{N}_{\text{star1}}\mathsf{N}_{\text{star1}}\mathsf{N}_{\text{star1}} = \mathsf{N}_{\text{star1}}, \tag{2.37}$$

$$\mathsf{N}_{\text{star2}}\mathsf{N}_{\text{star2}}\mathsf{N}_{\text{star2}} = \mathsf{N}_{\text{star2}}, \tag{2.38}$$

$$\mathsf{N}_{\text{circ1}}\mathsf{N}_{\text{circ1}}\mathsf{N}_{\text{circ1}}\mathsf{N}_{\text{circ1}}\mathsf{N}_{\text{circ1}} = \mathsf{N}_{\text{circ1}}, \tag{2.39}$$

$$\mathsf{N}_{\text{circ2}}\mathsf{N}_{\text{circ2}}\mathsf{N}_{\text{circ2}}\mathsf{N}_{\text{circ2}}\mathsf{N}_{\text{circ2}} = \mathsf{N}_{\text{circ2}}. \tag{2.40}$$

Note that there are no closed binary multiplications among the sets of 4-vertex matrices (2.35)–(2.36).

To give a proper group interpretation of (2.37)–(2.40), we introduce a *k-ary (polyadic) general linear semigroup* $\mathrm{GLS}^{[k]}(n, \mathbb{C}) = \{\mathsf{M}_{\mathrm{full}} | \mu^{[k]}\}$, where $\mathsf{M}_{\mathrm{full}} = \{M_{n\times n}\}$ is the set of $n \times n$ matrices over $\mathbb{C}$ and $\mu^{[k]}$ is an ordinary product of k matrices. The full semigroup $\mathrm{GLS}^{[k]}(n, \mathbb{C})$ is *derived* in the sense that its product can be obtained by repeating the binary products that are (binary) closed at each step.

Innovation 2.4. *The $n \times n$ matrices of special shape can form k-ary subsemigroups of* $\mathrm{GLS}^{[k]}(n, \mathbb{C})$ *that can be closed with respect to the product of at minimum k matrices but not of two matrices, and we call such semigroups k-ary nonderived (or k-nonderived).*

Moreover, we have for the sets $\mathsf{N}_{\mathrm{star1,\,2}}$ and $\mathsf{N}_{\mathrm{circ1,\,2}}$

$$\begin{aligned} &\mathsf{M}_{\mathrm{full}} = \mathsf{N}_{\mathrm{star1}} \bigcup \mathsf{N}_{\mathrm{star2}} \bigcup \mathsf{N}_{\mathrm{circ1}} \bigcup \mathsf{N}_{\mathrm{circ2}}, \\ &\mathsf{N}_{\mathrm{star1}} \bigcap \mathsf{N}_{\mathrm{star2}} \bigcap \mathsf{N}_{\mathrm{circ1}} \bigcap \mathsf{N}_{\mathrm{circ2}} = \varnothing. \end{aligned} \tag{2.41}$$

A simple example of a 3-nonderived subsemigroup of the full semigroup $\mathrm{GLS}^{[k]}(n, \mathbb{C})$ is the set of antidiagonal matrices $\mathsf{M}_{\mathrm{adiag}} = \left\{M_{\mathrm{adiag}}\right\}$ (having nonzero elements on the minor diagonal only): the product $\mu^{[3]}$ of three matrices from $\mathsf{M}_{\mathrm{adiag}}$ is closed, and therefore $\mathsf{M}_{\mathrm{adiag}}$ is a subsemigroup $\mathcal{S}^{[3]}_{\mathrm{adiag}} = \left\{\mathsf{M}_{\mathrm{adiag}} \middle| \mu^{[3]}\right\}$ of the full ternary general linear semigroup $\mathrm{GLS}^{[3]}(n, \mathbb{C})$ with the multiplication $\mu^{[3]}$ as the ordinary triple matrix product.

In the theory of polyadic groups (Dörnte 1929) an analog of the binary inverse M^{-1} is given by the *querelement*, which is denoted by $\bar{M}$ and in the matrix k-ary case is defined by

$$\overbrace{M \cdots M}^{k-1} \bar{M} = M, \tag{2.42}$$

where $\bar{M}$ can be on any place. If each element of the k-ary semigroup $\mathrm{GLS}^{[k]}(n, \mathbb{C})$ (or its subsemigroup) has its querelement $\bar{M}$, then this semigroup is a *k-ary general linear group* $\mathrm{GL}^{[k]}(n, \mathbb{C})$.

In the set of $n \times n$ matrices the binary (ordinary) product is defined (even it is not closed), and for invertible matrices we formally determine the standard inverse M^{-1}, but for arity $k \geqslant 4$ it does not coincide with the querelement $\bar{M}$ because, as follows from (2.42) and cancellativity in $\mathbb{C}$,

$$\bar{M} = M^{2-k}. \tag{2.43}$$

Definition 2.5. The *k-ary (polyadic) identity* $I_n^{[k]}$ in $\mathrm{GLS}^{[k]}(n, \mathbb{C})$ is defined by

$$\overbrace{I_n^{[k]} \cdots I_n^{[k]}}^{k-1} M = M, \tag{2.44}$$

which holds when M in the lhs is on any place.

If M is only on one or another side (but not in the middle places) in (2.44), then $I_n^{[k]}$ is called *left (right) polyadic identity*. For instance, in the subsemigroup (in $\mathrm{GLS}^{[k]}(n, \mathbb{C})$) of antidiagonal matrices $\mathcal{S}_{\mathrm{adiag}}^{[3]}$ the ternary identity $I_n^{[3]}$ can be chosen as the $n \times n$ reverse matrix (2.30) having units on the minor diagonal, while the ordinary $n \times n$ unit matrix I_n is not in $\mathcal{S}_{\mathrm{adiag}}^{[3]}$. It follows from (2.44), that for matrices over $\mathbb{C}$ the (left, right) polyadic identity $I_n^{[k]}$ is

$$\left(I_n^{[k]}\right)^{k-1} = I_n, \tag{2.45}$$

which means that for the ordinary matrix product $I_n^{[k]}$ is a $(k-1)$-root of I_n (or $I_n^{[k]}$ is a reflection of $(k-1)$ degree), while both sides cannot belong to a subsemigroup $\mathcal{S}^{[k]}$ of $\mathrm{GLS}^{[k]}(n, \mathbb{C})$ under consideration (as in $\mathcal{S}_{\mathrm{adiag}}^{[3]}$). Since the solutions of (2.45) are not unique, there can be many k-ary identities in a k-ary matrix semigroup. We denote the set of k-ary identities by $\mathsf{I}_n^{[k]} = \{I_n^{[k]}\}$. In the case of $\mathcal{S}_{\mathrm{adiag}}^{[3]}$ the ternary identity $I_n^{[3]}$ can be chosen as any of the $n \times n$ reverse matrices (2.30) with unit complex numbers $e^{i\alpha_j}$, $j = 1, \ldots, n$ on the minor diagonal, where α_j satisfies additional conditions depending on the semigroup. In the concrete case of $\mathcal{S}_{\mathrm{adiag}}^{[3]}$, the conditions giving (2.45) are $(k-1)\alpha_j = 1 + 2\pi r_j$, $r_j \in \mathbb{Z}$, $j = 1, \ldots, n$.

In the framework of the above definitions, we can interpret the closed products (2.37)–(2.38) as the multiplications $\mu^{[3]}$ of the *ternary semigroups* $\mathcal{S}_{\mathrm{star1,2}}^{[3]}(4, \mathbb{C}) = \{\mathsf{N}_{\mathrm{star1,2}} | \mu^{[3]}\}$. The corresponding querelements are given by

$$\bar{N}_{\mathrm{star1}} = N_{\mathrm{star1}}^{-1} = \begin{pmatrix} \frac{1}{x} & 0 & 0 & 0 \\ 0 & 0 & \frac{1}{z} & 0 \\ 0 & \frac{1}{y} & 0 & 0 \\ 0 & 0 & 0 & \frac{1}{t} \end{pmatrix},$$
$$\bar{N}_{\mathrm{star2}} = N_{\mathrm{star2}}^{-1} = \begin{pmatrix} 0 & 0 & 0 & \frac{1}{z} \\ 0 & \frac{1}{x} & 0 & 0 \\ 0 & 0 & \frac{1}{t} & 0 \\ \frac{1}{y} & 0 & 0 & 0 \end{pmatrix}, \quad x, y, z, t \neq 0. \tag{2.46}$$

The ternary semigroups having querelements for each element, i.e., the additional operation $\overline{(\,)}$ defined by (2.46), are the *ternary groups* $\mathcal{G}^{[3]}_{\text{star1, 2}}(4, \mathbb{C}) = \{\mathsf{N}_{\text{star1, 2}} | \mu^{[3]}, \overline{(\,)}\}$ which are two (non-intersecting because $\mathsf{N}_{\text{star1}} \cap \mathsf{N}_{\text{star2}} = \varnothing$) subgroups of the ternary general linear group $GL^{[3]}(4, \mathbb{C})$. The ternary identities in $\mathcal{G}^{[3]}_{\text{star1, 2}}(4, \mathbb{C})$ are the following different continuous sets $\mathsf{I}^{[3]}_{\text{star1, 2}} = \left\{I^{[3]}_{\text{star1, 2}}\right\}$, where

$$I^{[3]}_{\text{star1}} = \begin{pmatrix} e^{i\alpha_1} & 0 & 0 & 0 \\ 0 & 0 & e^{i\alpha_2} & 0 \\ 0 & e^{i\alpha_3} & 0 & 0 \\ 0 & 0 & 0 & e^{i\alpha_4} \end{pmatrix}, \quad e^{2i\alpha_1} = e^{2i\alpha_4} = e^{i(\alpha_2+\alpha_3)} = 1, \quad \alpha_j \in \mathbb{R}, \tag{2.47}$$

$$I^{[3]}_{\text{star2}} = \begin{pmatrix} 0 & 0 & 0 & e^{i\alpha_1} \\ 0 & e^{i\alpha_2} & 0 & 0 \\ 0 & 0 & e^{i\alpha_3} & 0 \\ e^{i\alpha_4} & 0 & 0 & 0 \end{pmatrix}, \quad e^{2i\alpha_2} = e^{2i\alpha_3} = e^{i(\alpha_1+\alpha_4)} = 1, \quad \alpha_j \in \mathbb{R}. \tag{2.48}$$

In the particular case $\alpha_j = 0$, $j = 1, 2, 3, 4$, the ternary identities (2.47)–(2.48) coincide with the bisymmetric permutation matrices (2.28).

Next we treat the closed set products (2.39)–(2.40) as the multiplications $\mu^{[5]}$ of the *5-ary semigroups* $\mathcal{S}^{[5]}_{\text{circ1, 2}}(4, \mathbb{C}) = \{\mathsf{N}_{\text{circ1, 2}} | \mu^{[5]}\}$. The querelements are

$$\bar{N}_{\text{circ1}} = N^{-3}_{\text{circ1}} = \begin{pmatrix} 0 & 0 & \frac{1}{yzt} & 0 \\ \frac{1}{xzt} & 0 & 0 & 0 \\ 0 & 0 & 0 & \frac{1}{xyt} \\ 0 & \frac{1}{xyz} & 0 & 0 \end{pmatrix}, \tag{2.49}$$

$$\bar{N}_{\text{circ2}} = N^{-3}_{\text{circ2}} = \begin{pmatrix} 0 & \frac{1}{yzt} & 0 & 0 \\ 0 & 0 & 0 & \frac{1}{xzt} \\ \frac{1}{xyt} & 0 & 0 & 0 \\ 0 & 0 & \frac{1}{xyz} & 0 \end{pmatrix}, \quad x, y, z, t \neq 0. \tag{2.50}$$

and the corresponding *5-ary groups* $\mathcal{G}^{[5]}_{\text{circ1, 2}}(4, \mathbb{C}) = \{\mathsf{N}_{\text{circ1, 2}} | \mu^{[5]}, \overline{(\,)}\}$, which are two (non-intersecting because $\mathsf{N}_{\text{circ1}} \cap \mathsf{N}_{\text{circ2}} = \varnothing$) subgroups of the 5-ary general linear group $\mathrm{GL}^{[5]}(n, \mathbb{C})$. We have the following continuous sets of 5-ary identities $\mathsf{I}^{[3]}_{\mathrm{circ}1,2} = \left\{ I^{[3]}_{\mathrm{circ}1,2} \right\}$ in $\mathscr{G}^{[5]}_{\mathrm{circ}1,2}(4, \mathbb{C})$ satisfying

$$I^{[5]}_{\mathrm{circ}1} = \begin{pmatrix} 0 & 0 & e^{i\alpha_1} & 0 \\ e^{i\alpha_2} & 0 & 0 & 0 \\ 0 & 0 & 0 & e^{i\alpha_3} \\ 0 & e^{i\alpha_4} & 0 & 0 \end{pmatrix}, \quad e^{i(\alpha_1+\alpha_2+\alpha_3+\alpha_4)} = 1, \quad \alpha_j \in \mathbb{R}, \tag{2.51}$$

$$I^{[5]}_{\mathrm{circ}2} = \begin{pmatrix} 0 & e^{i\alpha_1} & 0 & 0 \\ 0 & 0 & 0 & e^{i\alpha_2} \\ e^{i\alpha_3} & 0 & 0 & 0 \\ 0 & 0 & e^{i\alpha_4} & 0 \end{pmatrix}, \quad e^{i(\alpha_1+\alpha_2+\alpha_3+\alpha_4)} = 1, \quad \alpha_j \in \mathbb{R}. \tag{2.52}$$

In the case $\alpha_j = 0$, $j = 1, 2, 3, 4$, the 5-ary identities (2.51)–(2.52) coincide with the 90°-symmetric permutation matrices (2.29).

Innovation 2.6. *It follows from (2.46)–(2.52) that the 4-vertex star-like (2.35) and circle-like (2.36) matrices form subgroups of the k-ary general linear group* $\mathrm{GL}^{[k]}(4, \mathbb{C})$ *with significantly different properties: they have different querelements and (sets of) polyadic identities, and even the arities of the subgroups* $\mathscr{G}^{[3]}_{\mathrm{star}1,2}(4, \mathbb{C})$ *and* $\mathscr{G}^{[5]}_{\mathrm{circ}1,2}(4, \mathbb{C})$ *do not coincide (2.37)–(2.40).*

If we take into account that 4-vertex star-like (2.35) and circle-like (2.36) matrices are (binary) additive and distributive, then they form (with respect to the binary matrix addition (+) and the multiplications $\mu^{[3]}$ and $\mu^{[5]}$) the (2, 3)-ring $\mathscr{R}^{[3]}_{\mathrm{star}1,2}(4, \mathbb{C}) = \{\mathsf{N}_{\mathrm{star}1,2} | +, \mu^{[3]}\}$ and (2, 5)-ring $\mathscr{R}^{[5]}_{\mathrm{circ}1,2}(4, \mathbb{C}) = \{\mathsf{N}_{\mathrm{star}1,2} | +, \mu^{[5]}\}$.

Next we consider the interaction of the 4-vertex star-like (2.35) and circle-like (2.36) matrix sets, i.e., their exotic module structure. For this, let us recall the ternary (polyadic) module (Duplij 2001) and s-place action (Duplij 2018) definitions, which are suitable for our case. An abelian group $\mathscr{M}$ is a ternary left (middle, right) $\mathscr{R}$-module (or a module over $\mathscr{R}$) if there exists a ternary operation $\mathscr{R} \times \mathscr{R} \times \mathscr{M} \to \mathscr{M}$ ($\mathscr{R} \times \mathscr{M} \times \mathscr{R} \to \mathscr{M}$, $\mathscr{M} \times \mathscr{R} \times \mathscr{R} \to \mathscr{M}$) which satisfies some compatibility conditions (associativity and distributivity) that hold in the matrix case under consideration (and where the module operation is the triple ordinary matrix product) (Duplij 2001). A 5-ary left (right) module $\mathscr{M}$ over $\mathscr{R}$ is a 5-ary operation $\mathscr{R} \times \mathscr{R} \times \mathscr{R} \times \mathscr{R} \times \mathscr{M} \to \mathscr{M}$ ($\mathscr{M} \times \mathscr{R} \times \mathscr{R} \times \mathscr{R} \times \mathscr{R} \to \mathscr{M}$) with analogous conditions (and where the module operation is the pentuple matrix product) (Duplij 2018, 2022).

First, we have the triple relations inside star and circle matrices

$$\mathsf{N}_{\mathrm{star}1}(\mathsf{N}_{\mathrm{star}2})\mathsf{N}_{\mathrm{star}1} = (\mathsf{N}_{\mathrm{star}2}), \quad \mathsf{N}_{\mathrm{circ}1}\mathsf{N}_{\mathrm{circ}2}\mathsf{N}_{\mathrm{circ}1} = \mathsf{N}_{\mathrm{circ}1}, \tag{2.53}$$

$$\mathsf{N}_{\mathrm{star}1}\mathsf{N}_{\mathrm{star}1}(\mathsf{N}_{\mathrm{star}2}) = (\mathsf{N}_{\mathrm{star}2}), \quad \mathsf{N}_{\mathrm{circ}1}\mathsf{N}_{\mathrm{circ}1}\mathsf{N}_{\mathrm{circ}2} = \mathsf{N}_{\mathrm{circ}1}, \tag{2.54}$$

$$(\mathsf{N}_{\text{star2}})\mathsf{N}_{\text{star1}}\mathsf{N}_{\text{star1}} = (\mathsf{N}_{\text{star2}}), \quad \mathsf{N}_{\text{circ2}}\mathsf{N}_{\text{circ1}}\mathsf{N}_{\text{circ1}} = \mathsf{N}_{\text{circ1}}, \tag{2.55}$$

$$\mathsf{N}_{\text{star2}}\mathsf{N}_{\text{star2}}(\mathsf{N}_{\text{star1}}) = (\mathsf{N}_{\text{star1}}), \quad \mathsf{N}_{\text{circ2}}\mathsf{N}_{\text{circ2}}\mathsf{N}_{\text{circ1}} = \mathsf{N}_{\text{circ2}}, \tag{2.56}$$

$$\mathsf{N}_{\text{star2}}(\mathsf{N}_{\text{star1}})\mathsf{N}_{\text{star2}} = (\mathsf{N}_{\text{star1}}), \quad \mathsf{N}_{\text{circ2}}\mathsf{N}_{\text{circ1}}\mathsf{N}_{\text{circ2}} = \mathsf{N}_{\text{circ2}}, \tag{2.57}$$

$$(\mathsf{N}_{\text{star1}})\mathsf{N}_{\text{star2}}\mathsf{N}_{\text{star2}} = (\mathsf{N}_{\text{star1}}), \quad \mathsf{N}_{\text{circ1}}\mathsf{N}_{\text{circ2}}\mathsf{N}_{\text{circ2}} = \mathsf{N}_{\text{circ2}}. \tag{2.58}$$

We observe the following module structures on the left-hand column above (elements of the corresponding module are in brackets, and we informally denote modules by their sets): (1) from (2.53)–(2.55), the set $\mathsf{N}_{\text{star2}}$ is a middle, right, and left module over $\mathsf{N}_{\text{star1}}$; (2) from (2.56)–(2.58), the set $\mathsf{N}_{\text{star1}}$ is a middle, right, and left module over $\mathsf{N}_{\text{star2}}$;

$$\mathsf{N}_{\text{star1}}\mathsf{N}_{\text{circ1}}\mathsf{N}_{\text{star1}} = \mathsf{N}_{\text{circ2}}, \quad \mathsf{N}_{\text{star1}}\mathsf{N}_{\text{circ2}}\mathsf{N}_{\text{star1}} = \mathsf{N}_{\text{circ1}}, \tag{2.59}$$

$$\mathsf{N}_{\text{star2}}\mathsf{N}_{\text{circ1}}\mathsf{N}_{\text{star2}} = \mathsf{N}_{\text{circ2}}, \quad \mathsf{N}_{\text{star2}}\mathsf{N}_{\text{circ2}}\mathsf{N}_{\text{star2}} = \mathsf{N}_{\text{circ1}}, \tag{2.60}$$

$$\mathsf{N}_{\text{star1}}\mathsf{N}_{\text{star1}}(\mathsf{N}_{\text{circ1}}) = (\mathsf{N}_{\text{circ1}}), \quad (\mathsf{N}_{\text{circ1}})\mathsf{N}_{\text{star1}}\mathsf{N}_{\text{star1}} = (\mathsf{N}_{\text{circ1}}), \tag{2.61}$$

$$\mathsf{N}_{\text{star1}}\mathsf{N}_{\text{star1}}(\mathsf{N}_{\text{circ2}}) = (\mathsf{N}_{\text{circ2}}), \quad (\mathsf{N}_{\text{circ2}})\mathsf{N}_{\text{star1}}\mathsf{N}_{\text{star1}} = (\mathsf{N}_{\text{circ2}}), \tag{2.62}$$

$$\mathsf{N}_{\text{star2}}\mathsf{N}_{\text{star2}}(\mathsf{N}_{\text{circ1}}) = (\mathsf{N}_{\text{circ1}}), \quad (\mathsf{N}_{\text{circ1}})\mathsf{N}_{\text{star2}}\mathsf{N}_{\text{star2}} = (\mathsf{N}_{\text{circ1}}), \tag{2.63}$$

$$\mathsf{N}_{\text{star2}}\mathsf{N}_{\text{star2}}(\mathsf{N}_{\text{circ2}}) = (\mathsf{N}_{\text{circ2}}), \quad (\mathsf{N}_{\text{circ2}})\mathsf{N}_{\text{star2}}\mathsf{N}_{\text{star2}} = (\mathsf{N}_{\text{circ2}}), \tag{2.64}$$

(3) from (2.61)–(2.64), the sets $\mathsf{N}_{\text{circ1, 2}}$ are a right and left module over $\mathsf{N}_{\text{star1, 2}}$;

$$\mathsf{N}_{\text{circ1}}(\mathsf{N}_{\text{star1}})\mathsf{N}_{\text{circ1}} = (\mathsf{N}_{\text{star1}}), \quad \mathsf{N}_{\text{circ1}}(\mathsf{N}_{\text{star2}})\mathsf{N}_{\text{circ1}} = (\mathsf{N}_{\text{star2}}), \tag{2.65}$$

$$\mathsf{N}_{\text{circ2}}(\mathsf{N}_{\text{star1}})\mathsf{N}_{\text{circ2}} = (\mathsf{N}_{\text{star1}}), \quad \mathsf{N}_{\text{circ2}}(\mathsf{N}_{\text{star2}})\mathsf{N}_{\text{circ2}} = (\mathsf{N}_{\text{star2}}), \tag{2.66}$$

$$\mathsf{N}_{\text{circ1}}\mathsf{N}_{\text{circ1}}\mathsf{N}_{\text{star1}} = \mathsf{N}_{\text{star2}}, \quad \mathsf{N}_{\text{star1}}\mathsf{N}_{\text{circ1}}\mathsf{N}_{\text{circ1}} = \mathsf{N}_{\text{star2}}, \tag{2.67}$$

$$\mathsf{N}_{\text{circ1}}\mathsf{N}_{\text{circ1}}\mathsf{N}_{\text{star2}} = \mathsf{N}_{\text{star1}}, \quad \mathsf{N}_{\text{star2}}\mathsf{N}_{\text{circ1}}\mathsf{N}_{\text{circ1}} = \mathsf{N}_{\text{star1}}, \tag{2.68}$$

$$\mathsf{N}_{\text{circ2}}\mathsf{N}_{\text{circ2}}\mathsf{N}_{\text{star1}} = \mathsf{N}_{\text{star2}}, \quad \mathsf{N}_{\text{star1}}\mathsf{N}_{\text{circ2}}\mathsf{N}_{\text{circ2}} = \mathsf{N}_{\text{star2}}, \tag{2.69}$$

$$\mathsf{N}_{\text{circ2}}\mathsf{N}_{\text{circ2}}\mathsf{N}_{\text{star2}} = \mathsf{N}_{\text{star1}}, \quad \mathsf{N}_{\text{star2}}\mathsf{N}_{\text{circ2}}\mathsf{N}_{\text{circ2}} = \mathsf{N}_{\text{star1}}, \tag{2.70}$$

(4) from (2.65)–(2.66), the sets $\mathsf{N}_{\text{star1, 2}}$ are a middle ternary module over $\mathsf{N}_{\text{circ1, 2}}$;

$$\mathsf{N}_{\text{circ1}}\mathsf{N}_{\text{circ1}}\mathsf{N}_{\text{circ1}}\mathsf{N}_{\text{circ1}}(\mathsf{N}_{\text{star1}}) = (\mathsf{N}_{\text{star1}}), \quad \mathsf{N}_{\text{circ1}}\mathsf{N}_{\text{circ1}}\mathsf{N}_{\text{circ1}}\mathsf{N}_{\text{circ1}}(\mathsf{N}_{\text{star2}}) = (\mathsf{N}_{\text{star2}}), \tag{2.71}$$

$$(\mathsf{N}_{\text{star1}})\mathsf{N}_{\text{circ1}}\mathsf{N}_{\text{circ1}}\mathsf{N}_{\text{circ1}}\mathsf{N}_{\text{circ1}} = (\mathsf{N}_{\text{star1}}), \quad (\mathsf{N}_{\text{star2}})\mathsf{N}_{\text{circ1}}\mathsf{N}_{\text{circ1}}\mathsf{N}_{\text{circ1}}\mathsf{N}_{\text{circ1}} = (\mathsf{N}_{\text{star2}}), \tag{2.72}$$

$$\mathsf{N}_{\text{circ2}}\mathsf{N}_{\text{circ2}}\mathsf{N}_{\text{circ2}}\mathsf{N}_{\text{circ2}}(\mathsf{N}_{\text{star1}}) = (\mathsf{N}_{\text{star1}}), \quad \mathsf{N}_{\text{circ2}}\mathsf{N}_{\text{circ2}}\mathsf{N}_{\text{circ2}}\mathsf{N}_{\text{circ2}}(\mathsf{N}_{\text{star2}}) = (\mathsf{N}_{\text{star2}}), \tag{2.73}$$

$$(\mathsf{N}_{\mathrm{star1}})\mathsf{N}_{\mathrm{circ2}}\mathsf{N}_{\mathrm{circ2}}\mathsf{N}_{\mathrm{circ2}}\mathsf{N}_{\mathrm{circ2}} = (\mathsf{N}_{\mathrm{star1}}),\ (\mathsf{N}_{\mathrm{star2}})\mathsf{N}_{\mathrm{circ2}}\mathsf{N}_{\mathrm{circ2}}\mathsf{N}_{\mathrm{circ2}}\mathsf{N}_{\mathrm{circ2}} = (\mathsf{N}_{\mathrm{star2}}), \tag{2.74}$$

$$\mathsf{N}_{\mathrm{circ1}}\mathsf{N}_{\mathrm{circ1}}\mathsf{N}_{\mathrm{circ1}}\mathsf{N}_{\mathrm{circ1}}(\mathsf{N}_{\mathrm{circ2}}) = (\mathsf{N}_{\mathrm{circ2}}),\ (\mathsf{N}_{\mathrm{circ2}})\mathsf{N}_{\mathrm{circ1}}\mathsf{N}_{\mathrm{circ1}}\mathsf{N}_{\mathrm{circ1}}\mathsf{N}_{\mathrm{circ1}} = (\mathsf{N}_{\mathrm{circ2}}), \tag{2.75}$$

$$\mathsf{N}_{\mathrm{circ2}}\mathsf{N}_{\mathrm{circ2}}\mathsf{N}_{\mathrm{circ2}}\mathsf{N}_{\mathrm{circ2}}(\mathsf{N}_{\mathrm{circ1}}) = (\mathsf{N}_{\mathrm{circ1}}),\ (\mathsf{N}_{\mathrm{circ1}})\mathsf{N}_{\mathrm{circ2}}\mathsf{N}_{\mathrm{circ2}}\mathsf{N}_{\mathrm{circ2}}\mathsf{N}_{\mathrm{circ2}} = (\mathsf{N}_{\mathrm{circ1}}). \tag{2.76}$$

(5) from (2.71)–(2.76), the sets $\mathsf{N}_{\mathrm{circ1,\,2}}$ are right and left 5-ary modules over $\mathsf{N}_{\mathrm{circ2,\,1}}$ and $\mathsf{N}_{\mathrm{star1,\,2}}$.

Note that the sum of 4-vertex star solutions of the Yang–Baxter equations (2.33) (with different parameters) gives the shape of 8-vertex matrices, and the same with the 4-vertex circle solutions (2.34). Let us introduce two kind of 8-vertex 4×4 matrices over $\mathbb{C}$: an 8-vertex *star matrix* M_{star} and an 8-vertex *circle matrix* M_{circ} as

$$M_{\mathrm{star}} = \begin{pmatrix} x & 0 & 0 & y \\ 0 & z & s & 0 \\ 0 & t & u & 0 \\ v & 0 & 0 & w \end{pmatrix}, \quad \det M_{\mathrm{star}} = (xw - yv)(st - uz), \quad \operatorname{tr} M_{\mathrm{star}} = x + z + u + w, \tag{2.77}$$

$$M_{\mathrm{circ}} = \begin{pmatrix} 0 & x & y & 0 \\ z & 0 & 0 & s \\ t & 0 & 0 & u \\ 0 & v & w & 0 \end{pmatrix}, \quad \det M_{\mathrm{circ}} = (xw - yv)(st - uz), \quad \operatorname{tr} M_{\mathrm{circ}} = 0. \tag{2.78}$$

If M_{star} and M_{circ} are invertible (the determinants in (2.77)–(2.78) are non-vanishing), then

$$\begin{gathered} M_{\mathrm{star}}^{-1} = \begin{pmatrix} \frac{w}{xw - yv} & 0 & 0 & -\frac{v}{xw - yv} \\ 0 & -\frac{u}{st - uz} & \frac{t}{st - uz} & 0 \\ 0 & \frac{s}{st - uz} & -\frac{z}{st - uz} & 0 \\ -\frac{y}{xw - yv} & 0 & 0 & \frac{x}{xw - yv} \end{pmatrix}, \\ M_{\mathrm{circ}}^{-1} = \begin{pmatrix} 0 & \frac{w}{xw - yv} & -\frac{v}{xw - yv} & 0 \\ -\frac{u}{st - uz} & 0 & 0 & \frac{t}{st - uz} \\ \frac{s}{st - uz} & 0 & 0 & -\frac{z}{st - uz} \\ 0 & -\frac{y}{xw - yv} & \frac{x}{xw - yv} & 0 \end{pmatrix}, \end{gathered} \tag{2.79}$$

and therefore the parameter conditions for invertibility are the same in both M_{star} and M_{circ}

$$xw - yv \neq 0, \qquad st - uz \neq 0. \tag{2.80}$$

The corresponding sets $\mathsf{M}_{\text{star}} = \{M_{\text{star}}\}$ and $\mathsf{M}_{\text{circ}} = \{M_{\text{circ}}\}$ are closed under the following multiplications

$$\mathsf{M}_{\text{star}}\mathsf{M}_{\text{star}} = \mathsf{M}_{\text{star}}, \tag{2.81}$$

$$\mathsf{M}_{\text{star}}\mathsf{M}_{\text{circ}} = \mathsf{M}_{\text{circ}}, \qquad \mathsf{M}_{\text{circ}}\mathsf{M}_{\text{star}} = \mathsf{M}_{\text{circ}}, \tag{2.82}$$

$$\mathsf{M}_{\text{circ}}\mathsf{M}_{\text{circ}} = \mathsf{M}_{\text{star}}, \tag{2.83}$$

and in terms of sets we can write $\mathsf{M}_{\text{star}} = \mathsf{N}_{\text{star1}} \cup \mathsf{N}_{\text{star2}}$ and $\mathsf{M}_{\text{circ}} = \mathsf{N}_{\text{circ1}} \cup \mathsf{N}_{\text{circ2}}$, while $\mathsf{N}_{\text{star1}} \cap \mathsf{N}_{\text{star2}} = \varnothing$ and $\mathsf{N}_{\text{circ1}} \cap \mathsf{N}_{\text{circ2}} = \varnothing$ (see (2.41)). Note that, if M_{star} and M_{circ} are treated as elements of an algebra, then (2.81)–(2.83) are reminiscent of the Cartan decomposition (see, e.g., Helgason 1962), but we will consider them from a more general viewpoint, which will treat such structures as semigroups, ternary groups, and modules.

Innovation 2.7. *The set* $\mathsf{M}_{\text{8vertex}} = \mathsf{M}_{\text{star}} \cup \mathsf{M}_{\text{circ}}$ *is closed and because of the associativity of matrix multiplication,* $\mathsf{M}_{\text{8vertex}}$ *forms a non-commutative semigroup, which we call a 8-vertex matrix semigroup* $\mathcal{S}_{\text{8vertex}}(4, \mathbb{C})$*, which contains the zero matrix* $Z \in \mathcal{S}_{\text{8vertex}}(4, \mathbb{C})$ *and is a subsemigroup of the (binary) general linear semigroup* $\mathrm{GLS}(4, \mathbb{C})$.

It follows from (2.81), that M_{star} is its subsemigroup $\mathcal{S}^{\text{star}}_{\text{8vertex}}(4, \mathbb{C})$. Moreover, the invertible elements of $\mathcal{S}_{\text{8vertex}}(4, \mathbb{C})$ form a *8-vertex matrix group* $\mathcal{G}_{\text{8vertex}}(4, \mathbb{C})$ because its identity is a unit 4×4 matrix $I_4 \in \mathsf{M}_{\text{star}}$, and so M_{star} is a subgroup $\mathcal{G}^{\text{star}}_{\text{8vertex}}(4, \mathbb{C})$ of $\mathcal{G}_{\text{8vertex}}(4, \mathbb{C})$ and a subgroup of the (binary) general linear group $\mathrm{GL}(4, \mathbb{C})$. The structure of $\mathcal{S}_{\text{8vertex}}(4, \mathbb{C})$ (2.81) is similar to that of block-diagonal and block-antidiagonal matrices (of the necessary sizes). So the 8-vertex (binary) matrix semigroup $S_{\text{8vertex}}(4, \mathbb{C})$ in which the parameters satisfy (2.80) is a 8-vertex (binary) matrix group $\mathcal{G}_{\text{8vertex}}(4, \mathbb{C})$, having a subgroup $\mathcal{G}^{\text{star}}_{\text{8vertex}}(4, \mathbb{C}) = \langle \mathsf{M}_{\text{star}} | \cdot, \mathcal{I}_4 \rangle$, where $(\cdot)$ is an ordinary matrix product, and I_4 is its identity.

The group structure of the circle matrices M_{circ} (2.78) follows from

$$\mathsf{M}_{\text{circ}}\mathsf{M}_{\text{circ}}\mathsf{M}_{\text{circ}} = \mathsf{M}_{\text{circ}}, \tag{2.84}$$

which means that M_{circ} is closed with respect to the product of three matrices (the product of two matrices from M_{circ} is outside the set (2.83)). We define a ternary multiplication $\nu^{[3]}$ as the ordinary triple product of matrices.

Innovation 2.8. *Then* $\mathcal{S}^{\text{circ}[3]}_{\text{8vertex}}(4, \mathbb{C}) = \langle \mathsf{M}_{\text{circ}} | \nu^{[3]} \rangle$ *becomes a ternary (3-nonderived) semigroup with the zero* $Z \in \mathsf{M}_{\text{circ}}$*, which is a subsemigroup of the ternary (derived) general linear semigroup* $\mathrm{GLS}^{[3]}(4, \mathbb{C})$*. Instead of the inverse, for each invertible element* $M_{\text{circ}} \in \mathsf{M}_{\text{circ}} \backslash Z$ *we introduce the unique querelement* $\bar{M}_{\text{circ}}$ (Dörnte 1929) *by*

(2.42), *and because the ternary product is the triple ordinary product, we have* $\bar{M}_{\text{circ}} = M_{\text{circ}}^{-1}$ *from (2.43).*

Innovation 2.9. *If the conditions of invertibility (2.80) hold valid, then the ternary semigroup* $\mathcal{S}_{8\text{vertex}}^{\text{circ}(3)}(4, \mathbb{C})$ *becomes the ternary group* $\mathcal{G}_{8\text{vertex}}^{\text{circ}(3)}(4, \mathbb{C}) = \langle \mathsf{M}_{\text{circ}} | \nu^{[3]}, \bar{()} \rangle$ *which does not contain the ordinary (binary) identity, since* $I_4 \notin \mathsf{M}_{\text{circ}}$.

Nevertheless, the ternary group of circle matrices $\mathcal{G}_{8\text{vertex}}^{\text{circ}[3]}(4, \mathbb{C})$ has the following set $\mathsf{I}_{\text{circ}}^{[3]} = \left\{ I_{\text{circ}}^{[3]} \right\}$ of left–right 6-vertex and 8-vertex ternary identities (see (2.44)–(2.45))

$$I_{\text{circ}}^{[3]} = \begin{pmatrix} 0 & \frac{1}{a} & b & 0 \\ a & 0 & 0 & -\frac{ab}{c} \\ 0 & 0 & 0 & \frac{1}{c} \\ 0 & 0 & c & 0 \end{pmatrix}, \begin{pmatrix} 0 & -\frac{ab}{c} & \frac{1}{c} & 0 \\ 0 & 0 & 0 & \frac{1}{b} \\ c & 0 & 0 & a \\ 0 & b & 0 & 0 \end{pmatrix}, \begin{pmatrix} 0 & -\frac{ab}{c} & \frac{1-ad}{c} & 0 \\ -\frac{cd}{b} & 0 & 0 & \frac{1-ad}{b} \\ c & 0 & 0 & a \\ 0 & b & d & 0 \end{pmatrix}, \tag{2.85}$$

which (without additional conditions) depend upon the free parameters $a, b, c, d \in \mathbb{C}$, $b, c \neq 0$, and $\left(I_{\text{circ}}^{[3]}\right)^2 = I_4$, $I_{\text{circ}}^{[3]} \in \mathsf{M}_{\text{circ}}$. In the binary sense, the matrices from (2.85) are mutually similar, but as ternary identities they are different.

If we consider the second operation for matrices (as elements of a general matrix ring), the binary matrix addition (+), then he structure of $\mathsf{M}_{8\text{vertex}} = \mathsf{M}_{\text{star}} \cup \mathsf{M}_{\text{circ}}$ becomes more exotic.

Innovation 2.10. *The set* M_{star} *is a* (2, 2)*-ring* $\mathcal{R}_{8\text{vertex}}^{star[2,2]} = \langle \mathsf{M}_{\text{star}} | +, \cdot \rangle$ *with the binary addition* (+) *and binary multiplication* $(\cdot)$ *from the semigroup* $\mathcal{S}_{8\text{vertex}}^{\text{star}}$, *while* M_{circ} *is a* (2, 3)*-ring* $\mathcal{R}_{8\text{vertex}}^{\text{circ}[2,3]} = \langle \mathsf{M}_{\text{circ}} | +, \nu^{[3]} \rangle$ *with the binary matrix addition* (+), *the ternary matrix multiplication* $\nu^{[3]}$ *and the zero* Z.

Moreover, because of the distributivity and associativity of binary matrix multiplication, the relations (2.82) mean that the set M_{circ} (being an abelian group under binary addition) can be treated as a left and right binary module $\mathcal{M}_{8\text{vertex}}^{\text{circ}}$ over the ring $\mathcal{R}_{8\text{vertex}}^{star(2,2)}$ with an operation (*): the module action $M_{\text{circ}} * M_{\text{star}} = M_{\text{circ}}$, $M_{\text{star}} * M_{\text{circ}} = M_{\text{circ}}$ (coinciding with the ordinary matrix product (2.82)). The left and right modules are compatible because the associativity of ordinary matrix multiplication gives the compatibility condition $(M_{\text{circ}} M_{\text{star}}) M'_{\text{circ}} = M_{\text{circ}} (M_{\text{star}} M'_{\text{circ}})$, $M_{\text{star}} \in \mathcal{R}_{8\text{vertex}}^{star(2,2)}$, $M_{\text{circ}}, M'_{\text{circ}} \in \mathcal{R}_{8\text{vertex}}^{\text{circ}(2,3)}$, and therefore M_{circ} (as an abelian group under the binary addition (+) and the module action (*)) is a $\mathcal{R}_{8\text{vertex}}^{star(2,2)}$-bimodule $\mathcal{M}_{8\text{vertex}}^{\text{circ}}$. The last relation (2.83) shows another interpretation of M_{circ} as a formal square root of M_{star} (as sets).

2.1.6 Star 8-vertex and circle 8-vertex Yang–Baxter maps

Let us consider the star 8-vertex solutions $\tilde{c}$ to the Yang–Baxter equation (2.12), having the shape (2.77), in the most general setting, over $\mathbb{C}$ and for different ranks, i.e., including noninvertible ones. In components, they are determined by

$$\begin{aligned}
&vy(u-z)=0,\quad y(t^2-wz-x^2+xz)=0,\quad y(s(x-z)+t(u-x))=0,\\
&y(u(w-x)+x^2-s^2)=0,\quad svy-tuz=0,\quad tvy-suz=0,\\
&vwy+xz(x-z)-stz=0,\quad y(w^2-wz+xz-s^2)=0,\\
&uz(z-u)=0,\quad suz-tvy=0,\quad y(s(w-u)+t(z-w))=0,\\
&tuz-svy=0,\quad stz-vxy+wz(z-w)=0,\quad v(s^2-wz-x^2+xz)=0,\\
&stu+u^2x-ux^2-vwy=0,\quad v(s(z-w)+t(w-u))=0,\\
&uz(u-z)=0,\quad y(t^2+u(w-x)-w^2)=0,\\
&v(s(u-x)+t(x-z))=0,\quad v(s^2+u(w-x)-w^2)=0,\\
&vy(z-u)=0,\quad v(u(w-x)+x^2-t^2)=0,\\
&uw^2+vxy-stu-u^2w=0,\quad v(w^2-t^2-wz+xz)=0.
\end{aligned}\tag{2.86}$$

Solutions from, e.g., Dye (2003) and Hietarinta (1993), etc, should satisfy this overdetermined system of 24 cubic equations for eight variables.

We search for the 8-vertex constant solutions to the Yang–Baxter equation over $\mathbb{C}$ without additional conditions, unitarity, etc (which will be considered in the next sections). We also will need the matrix functions tr and det, which are related to link invariants, as well as the eigenvalues, which help to find similar matrices and q-conjugated solutions to braid equations. Take into account that the Yang–Baxter maps are determined up to a general complex factor $t \in \mathbb{C}$ (2.14). For eigenvalues (which are determined up to the same factor t) we use the notation: $\{\text{eigenvalue}\}^{[\text{algebraic multiplicity}]}$.

We found the following 8-vertex solutions, classified by rank and number of parameters.

- Rank = 4 (invertible star Yang–Baxter maps) are

(1) Quadratic in two parameters

$$\tilde{c}^{\,\text{par}=2}_{\text{rank}=4}(x,y)=\begin{pmatrix} xy & 0 & 0 & y^2\\ 0 & xy & \pm xy & 0\\ 0 & \mp xy & xy & 0\\ -x^2 & 0 & 0 & xy \end{pmatrix},\tag{2.87}$$

$$\operatorname{tr}\tilde{c}=4xy,$$

$$\det\tilde{c}=4x^4y^4,\quad x\neq 0,\ y\neq 0,$$

$$\text{eigenvalues: } \{(1+i)xy\}^{[2]},\ \{(1-i)xy\}^{[2]},$$

(2) Quadratic in three parameters

$$\tilde{c}^{\text{par}=3}_{\text{rank}=4,\,1}(x, y, z) = \begin{pmatrix} xy & 0 & 0 & y^2 \\ 0 & zy & \pm xy & 0 \\ 0 & \pm xy & zy & 0 \\ z^2 & 0 & 0 & xy \end{pmatrix}, \tag{2.88}$$

$$\operatorname{tr} \tilde{c} = 2y(x + z),$$
$$\det \tilde{c} = y^4(z^2 - x^2)^2, \quad z \neq \pm x, \quad y \neq 0,$$
$$\text{eigenvalues:} y(x - z), -y(x - z), \{y(x + z)\}^{[2]},$$

(3) Irrational in three parameters

$$\tilde{c}^{\text{par}=3}_{\text{rank}=4}(x, y, z) = \begin{pmatrix} xy & 0 & 0 & y^2 \\ 0 & \frac{x+z}{2}y & \pm y\sqrt{\frac{x^2+z^2}{2}} & 0 \\ 0 & \pm y\sqrt{\frac{x^2+z^2}{2}} & \frac{x+z}{2}y & 0 \\ \frac{(x+z)^2}{4} & 0 & 0 & yz \end{pmatrix}, \tag{2.89}$$

$$\operatorname{tr} \tilde{c} = 2y(x + z),$$
$$\det \tilde{c} = \frac{1}{16}y^4(x - z)^4, \quad y \neq 0, z \neq x,$$
$$\text{eigenvalues} \quad : \left\{\frac{1}{2}y\left(x + z - \sqrt{2}\sqrt{x^2 + z^2}\right)\right\}^{[2]}, \left\{\frac{1}{2}y\left(x + z + \sqrt{2}\sqrt{x^2 + z^2}\right)\right\}^{[2]}.$$

Note that only the first and the last cases are genuine 8-vertex Yang–Baxter maps because the three-parameter matrices (2.88) are q-conjugated with the 4-vertex parameter-permutation solutions (2.33). Indeed,

$$\begin{pmatrix} xy & 0 & 0 & y^2 \\ 0 & zy & xy & 0 \\ 0 & xy & zy & 0 \\ z^2 & 0 & 0 & xy \end{pmatrix} = (q \otimes_K q)\begin{pmatrix} y(x+z) & 0 & 0 & 0 \\ 0 & 0 & y(x-z) & 0 \\ 0 & y(x-z) & 0 & 0 \\ 0 & 0 & 0 & y(x+z) \end{pmatrix}(q^{-1} \otimes_K q^{-1}), \tag{2.90}$$

$$q = \begin{pmatrix} \pm\sqrt{\frac{y}{z}} & b \\ 1 & \mp b\sqrt{\frac{z}{y}} \end{pmatrix}, \tag{2.91}$$

where $b \in \mathbb{C}$ is a free parameter. If $b = \frac{y}{z}$ two matrices q in (2.91) are similar, and we have the unique q-conjugation (2.90), then another solution in (2.88) is q-conjugated to the second 4-vertex parameter-permutation solutions (2.33) such that

$$\begin{pmatrix} xy & 0 & 0 & y^2 \\ 0 & zy & -xy & 0 \\ 0 & -xy & zy & 0 \\ z^2 & 0 & 0 & xy \end{pmatrix} = (q \otimes_{\mathrm{K}} q) \begin{pmatrix} 0 & 0 & 0 & y(x-z) \\ 0 & y(x+z) & 0 & 0 \\ 0 & 0 & y(x+z) & 0 \\ y(x-z) & 0 & 0 & 0 \end{pmatrix} (q^{-1} \otimes_{\mathrm{K}} q^{-1}), \tag{2.92}$$

$$q = \begin{pmatrix} i\sqrt{\frac{y}{z}} & \pm i\sqrt{\frac{y}{z}} \\ \pm 1 & 1 \end{pmatrix}, \begin{pmatrix} -i\sqrt{\frac{y}{z}} & \pm i\sqrt{\frac{y}{z}} \\ \pm 1 & 1 \end{pmatrix}, \tag{2.93}$$

where qs are pairwise similar in (2.93), and therefore we have two different q-conjugations.

- Rank = 2 (noninvertible star Yang–Baxter maps) are quadratic in parameters

$$\tilde{c}^{\,\mathrm{par}=2}_{\mathrm{rank}=2}(x, y) = \begin{pmatrix} xy & 0 & 0 & y^2 \\ 0 & xy & \pm xy & 0 \\ 0 & \pm xy & xy & 0 \\ x^2 & 0 & 0 & xy \end{pmatrix}, \quad \begin{array}{c} \mathrm{tr}\, \tilde{c} = 4xy, \\ \text{eigenvalues:} \{2xy\}^{[2]}, \{0\}^{[2]}. \end{array} \tag{2.94}$$

There are no star 8-vertex solutions of rank 3. The above two solutions for $\tilde{c}^{\,\mathrm{par}=2}_{\mathrm{rank}=4}$ with different signs are q-conjugated (2.19), with the matrix q being one of the following

$$q = \begin{pmatrix} 0 & 1 \\ \pm i\frac{x}{y} & 0 \end{pmatrix}. \tag{2.95}$$

Further families of solutions can be obtained from (2.87)–(2.94) by applying the general q-conjugation (2.14).

Particular cases of the star solutions are called also X-type operators (Padmanabhan *et al* 2021) or magic matrices (Ballard and Wu 2011b) connected with the Cartan decomposition of $SU(4)$ (Khaneja and Glaser 2001, Kraus and Cirac 2001, Bullock 2004, Bullock and Brennen 2004).

The circle 8-vertex solutions $\tilde{c}$ to the Yang–Baxter equation (2.12) of the shape (2.78) are determined by the following system of 32 cubic equations for eight unknowns over $\mathbb{C}$

$$\begin{aligned}
&x(ty + z(u - y) - vx) = 0,\ tx^2 + y^2(v - z) - wx^2 = 0,\\
&y(-st + tx + wy - xz) = 0,\ su(x - y) - sxy + uxy = 0,\\
&z(t(y - x) - sz + wx) = 0,\ v(sy + x^2) - z(s^2 + ux) = 0,\\
&swy - s^2v + xy(v - z) = 0,\ swx - s^2w + yz(u - y) = 0,\\
&st^2 - t^2x + z^2(y - u) = 0,\ su(v - z) + x(xz - tu) = 0,\\
&su(w - v) + xy(z - t) = 0,\ s(tu - uw + yz) - ty^2 = 0,\\
&s(tv + z^2) - x(v^2 + wz) = 0,\ svw - vwx + z(xz - wy) = 0,\\
&sw(w - t) + yz(z - v) = 0,\ s(sz + u(v - w) - vy) = 0,\\
&t(tu - vy + z(y - x)) = 0,\ tx(x - s) + u^2v - uvy = 0,\\
&xy(t - w) + u^2w - uvx = 0,\ t(sy + u^2) - w(ux + y^2) = 0,\\
&tz(s - x) - sv^2 + tuv = 0,\ tz(x - y) - svw + uvw = 0,\\
&u(w^2 - tz) - swz + tyz = 0,\ s^2(t - w) + u^2(v - z) = 0,\\
&tx(w - t) + uv(z - v) = 0,\ tvx - t^2y - uvw + vwy = 0,\\
&tvy - t^2u + w(wy - uz) = 0,\ u(s(v - w) - tu + wx) = 0,\\
&twz - tv(w + z) + vwz = 0,\ v(s(w - t) - uw + vx) = 0,\\
&sw^2 - uv^2 + v^2y - w^2x = 0,\ w(sv + u(z - v) - wy) = 0.
\end{aligned} \tag{2.96}$$

We found the 8-vertex solutions, classified by rank and number of parameters.

• Rank = 4 (invertible circle Yang–Baxter map) are quadratic in parameters

$$\begin{gathered}
\tilde{c}^{\,\text{par}=3}_{\text{rank}=4}(x, y, z) = \begin{pmatrix} 0 & xy & yz & 0 \\ z^2 & 0 & 0 & xy \\ xz & 0 & 0 & yz \\ 0 & z^2 & xz & 0 \end{pmatrix},\\
\operatorname{tr} \tilde{c} = 0,\\
\det \tilde{c} = y^2z^2(z^2 - x^2),\ y \neq 0,\ \ z \neq 0,\ z \neq \pm x,\\
\text{eigenvalues:} \sqrt{-yz}(x - z),\ -\sqrt{-yz}(x - z),\ \sqrt{yz}(x + z),\ -\sqrt{yz}(x + z).
\end{gathered} \tag{2.97}$$

• Rank = 2 (noninvertible circle Yang–Baxter map) are linear in parameters

$$\tilde{c}^{\,\text{par}=2}_{\text{rank}=2}(x, y) = \begin{pmatrix} 0 & -y & -y & 0 \\ -x & 0 & 0 & y \\ -x & 0 & 0 & y \\ 0 & x & x & 0 \end{pmatrix},\ \ \text{eigenvalues: } 2\sqrt{xy},\ -2\sqrt{xy},\ \{0\}^{[2]}. \tag{2.98}$$

There are no circle 8-vertex solutions of rank 3. The corresponding families of solutions can be derived from the above using the q-conjugation (2.14).

A particular case of the 8-vertex circle solution (2.97) was considered in Asaulko and Korablev (2019).

2.1.7 Triangle invertible 9- and 10-vertex solutions

There are some higher vertex solutions to the Yang–Baxter equations that are not in the above star/circle classification. They are determined by the following system of 15 cubic equations for nine unknowns over $\mathbb{C}$

$$
\begin{aligned}
&(-py - x(u + w - y) + v(y + z)) + s(v - x)(v + x) = 0,\\
&(-ty + vz + x(y - z)) = 0,\\
&x(t - v) + ty(w - z) + vz(y - u) = 0,\\
&(pz - t(y + z) + x(u + w - z)) + s(x^2 - t^2) = 0,\\
&s(-u + w - y + z) + s(-t(u + z) + x(u - w + y - z) + v(w + y))\\
&uwy - uwz - uyz + wyz = 0,\\
&y(p - t) + u(x - v)) = 0,\ t(pz - t(u + z) + ux) = 0,\\
&{}^2 s + pu(-u + y + z) - t(st + u(u + w)) + u^2 x = 0,\ t(z(p - v) - tw + wx) = 0,\\
&s(t - v) + tw(y - u) + uv(w - z) = 0,\ v(-py + v(w + y) - wx) = 0,\\
&(z(v - p) + tw - wx) = 0,\ v(y(t - p) + u(v - x)) = 0,\\
&{}^2(-s) + pw(w - y - z) + sv^2 + w(v(u + w) - wx) = 0,\\
&(p(w - u) - tw + uv) = 0,
\end{aligned}
\tag{2.99}
$$

We found the following 9-vertex Yang–Baxter maps

$$
\tilde{c}^{\,9-\text{vert},\,1}_{\text{rank}=4} = \begin{pmatrix} x & y & z & s \\ 0 & 0 & x & y \\ 0 & x & 0 & z \\ 0 & 0 & 0 & x \end{pmatrix}, \begin{pmatrix} x & y & y & z \\ 0 & 0 & -x & -y \\ 0 & -x & 0 & -y \\ 0 & 0 & 0 & x \end{pmatrix}, \begin{pmatrix} x & y & -y & z \\ 0 & 0 & x & -\dfrac{zx}{y} \\ 0 & x & 0 & \dfrac{zx}{y} \\ 0 & 0 & 0 & x \end{pmatrix}, \tag{2.100}
$$

$$
\operatorname{tr} \tilde{c} = 2x, \quad \det \tilde{c} = -x^4, \quad x \neq 0, \quad \text{eigenvalues: } \{x\}^{[3]},\ -x. \tag{2.101}
$$

The third matrix in (2.100) is conjugated with the 4-vertex parameter-permutation solutions (2.33) of the form (which has the same the same eigenvalues (2.101))

$$
\tilde{c}^{\,4-\text{vert}}_{\text{rank}=4}(x) = \begin{pmatrix} x & 0 & 0 & 0 \\ 0 & 0 & x & 0 \\ 0 & x & 0 & 0 \\ 0 & 0 & 0 & x \end{pmatrix} \tag{2.102}
$$

by the conjugated matrix

$$
U^{9\text{to}4} = \begin{pmatrix} 1 & -\dfrac{y}{2x} & \dfrac{y}{2x} & 0 \\ 0 & 1 & 0 & -\dfrac{z}{y} \\ 0 & 0 & 1 & 0 \\ 0 & 0 & 0 & 1 \end{pmatrix}. \tag{2.103}
$$

The matrix (2.103) cannot be presented as the Kronecker product $q \otimes_{\mathrm{K}} q$ (2.16), and so the third matrix in (2.100) and (2.102) are different solutions of the Yang–Baxter equation (2.12). Although the first two matrices in (2.100) have the same eigenvalues (2.101), they are not similar because they are different from (2.102) middle Jordan blocks.

Then we have another 3-parameter solutions with fractions

$$\tilde{c}_{\mathrm{rank}=4}^{9-\mathrm{vert},\,2}(x, y, z) = \begin{pmatrix} x & y & y & z \\ 0 & 0 & -x & y - \dfrac{2xz}{y} \\ 0 & -x & 0 & y - \dfrac{2xz}{y} \\ 0 & 0 & 0 & x\left(\dfrac{4xz}{y^2} - 3\right) \end{pmatrix}, \tag{2.104}$$

$$\operatorname{tr} \tilde{c} = 2x\frac{2xz - y^2}{y^2},$$

$$\det \tilde{c} = x^4\left(3 - \frac{4xz}{y^2}\right), \quad x \neq 0, \ y \neq 0, \ z \neq \frac{3y^2}{4x},$$

$$\text{eigenvalues: } \{x\}^{[2]}, -x, x\left(\frac{4xz}{y^2} - 3\right),$$

and

$$\tilde{c}_{\mathrm{rank}=4}^{9-\mathrm{vert},\,3}(x, y, z) = \begin{pmatrix} x & y & -y & z \\ 0 & 0 & -x & \dfrac{2zx}{y} + y \\ 0 & 3x & 0 & \dfrac{2zx}{y} - y \\ 0 & 0 & 0 & \dfrac{4zx^2}{y^2} + x \end{pmatrix}, \tag{2.105}$$

$$\operatorname{tr} \tilde{c} = 2x\left(1 + 2\frac{xz}{y^2}\right)$$

$$\det \tilde{c} = 3x^4\left(\frac{4zx}{y^2} + 1\right), \quad x \neq 0, \ y \neq 0, \ z \neq \frac{y^2}{4x}$$

$$\text{eigenvalues: } x, i\sqrt{3}x, -i\sqrt{3}x, x\left(1 + \frac{4zx}{y^2}\right)$$

The 4-parameter 9-vertex solution is

$$
{}^{9-\text{vert,par}=4}_{\text{rank}=4}(x, y, z, s) = \begin{pmatrix} x & y & z & s \\ 0 & 0 & -x & y - \dfrac{2sx}{z} \\ 0 & x - \dfrac{2xy}{z} & 0 & z - \dfrac{2sx}{z} \\ 0 & 0 & 0 & \dfrac{x(4sx - z(2y + z))}{z^2} \end{pmatrix},
$$

$$
\text{tr}\,\tilde{c} = 2x\frac{2sx - yz}{z^2}
$$

$$
\det \tilde{c} = \frac{x^4(2y - z)(z(2y + z) - 4sx)}{z^3}, \; x \neq 0, \; y \neq \frac{z}{2}, \; z \neq 0,
$$

$$
\text{eigenvalues}: \; x, \; x\sqrt{\frac{2y}{z} - 1}, \; -x\sqrt{\frac{2y}{z} - 1}, \; \frac{x(4sx - z(2y + z))}{z^2}. \tag{2.106}
$$

We also found a 5-parameter and 9-vertex solution of the form

$$
\tilde{c}^{9-\text{vert,par}=5}_{\text{rank}=4}(x, y, z, s, t) = \begin{pmatrix} x & y & z & s \\ 0 & 0 & t & \dfrac{s(t - x)}{z} + y \\ 0 & \dfrac{y(t - x)}{z} + x & 0 & \dfrac{s(t - x)}{z} + z \\ 0 & 0 & 0 & \dfrac{s(t - x)^2 + tz(y + z) - xyz}{z^2} \end{pmatrix},
$$

$$
\text{tr}\,\tilde{c} = \frac{st^2 + sx^2 + tz^2 + xz^2 - 2stx + tyz - xyz}{z^2},
$$

$$
\det \tilde{c} = \frac{xt(x(y - z) - ty)(s(t - x)^2 + tz(y + z) - xyz)}{z^3},
$$

$$
\text{eigenvalues}: \; x, \; \sqrt{\frac{t}{z}(ty - xy + xz)}, \; -\sqrt{\frac{t}{z}(ty - xy + xz)},
$$

$$
\frac{st^2 - 2stx + tz^2 + ytz + sx^2 - yxz}{z^2}, \; x \neq 0, \; z \neq 0, \; t \neq 0. \tag{2.107}
$$

Finally, we found the following 3-parameter 10-vertex solution

$$
\tilde{c}^{10-\text{vert}}_{\text{rank}=4}(x, y, z) = \begin{pmatrix} x & y & y & \dfrac{y^2}{x} \\ 0 & 0 & -x & -y \\ 0 & -x & 0 & -y \\ z & 0 & 0 & x \end{pmatrix},
$$

$$
\text{tr}\,\tilde{c} = 2x,
$$

$$
\det \tilde{c} = -x(x^3 + zy^2), \;\; x \neq 0,
$$

$$
\text{eigenvalues:} \; \{x\}^{[2]}, \; \sqrt{x^2 + \frac{zy^2}{x}}, \; -\sqrt{x^2 + \frac{zy^2}{x}}. \tag{2.108}
$$

This solution is conjugated with the 4-vertex parameter-permutation solutions (2.33) of the form (which has the same the same eigenvalues as (2.108))

$$\tilde{c}^{\,4-\mathrm{vert}}_{\mathrm{rank}=4}(x, y, z) = \begin{pmatrix} x & 0 & 0 & 0 \\ 0 & 0 & x + \dfrac{y^2 z}{x^2} & 0 \\ 0 & x & 0 & 0 \\ 0 & 0 & 0 & x \end{pmatrix} \tag{2.109}$$

by the conjugated matrix

$$U^{10\mathrm{to}4} = \begin{pmatrix} 0 & \dfrac{x}{z} & -\dfrac{x}{z} & 0 \\ -1 & -\dfrac{x^2}{yz} & \dfrac{x^2}{yz} & -\dfrac{y}{x} \\ 1 & -\dfrac{x^2}{yz} & \dfrac{x^2}{yz} & 0 \\ 0 & 0 & 1 & 1 \end{pmatrix}. \tag{2.110}$$

Since the matrix (2.110) cannot be presented as the Kronecker product $q \otimes_{\mathrm{K}} q$ (2.16), (2.108) and (2.109) are different solutions of the Yang–Baxter equation (2.12).

Further families of the higher vertex solutions to the constant Yang–Baxter equation (2.12) can be obtained from the ones above by using the q-conjugation (2.14).

2.2 Polyadic braid operators and higher braid equations

The polyadic version of the braid equation (2.1) was introduced in Duplij (2021b, 2022). Here we define higher analog of the Yang–Baxter operator and develop its connection with higher braid groups and quantum computations. The whole material of this and the following sections is fully original and innovative.

Let us consider a vector space $\mathcal{V}$ over a field $\mathbb{K}$. A *polyadic (n-ary) braid operator* $C_{\mathcal{V}^n}$ is defined as the mapping (Duplij 2021b)

$$C_{\mathcal{V}^n}\colon \overbrace{\mathcal{V}\otimes \cdots \otimes \mathcal{V}}^{n} \to \overbrace{\mathcal{V}\otimes \cdots \otimes \mathcal{V}}^{n}. \tag{2.111}$$

The polyadic analog of the braid equation (2.1) was introduced in Duplij (2021b) using the associative quiver technique (Duplij 2018).

Let us introduce n operators

$$A_p\colon \overbrace{\mathcal{V}\otimes \cdots \otimes \mathcal{V}}^{2n-1} \to \overbrace{\mathcal{V}\otimes \cdots \otimes \mathcal{V}}^{2n-1}, \tag{2.112}$$

$$A_p = \mathrm{id}_{\mathcal{V}}^{\otimes(p-1)} \otimes C_{\mathcal{V}^n} \otimes \mathrm{id}_{\mathcal{V}}^{\otimes(n-p)}, \quad p = 1, \ldots, n, \tag{2.113}$$

i.e., p is a place of $C_{\mathcal{V}^n}$ instead of one $\mathrm{id}_{\mathcal{V}}$ in $\mathrm{id}_{\mathcal{V}}^{\otimes n}$. A system of $(n-1)$ *polyadic (n-ary) braid equations* is defined by

$$A_1 \circ A_2 \circ A_3 \circ A_4 \circ \cdots \circ A_{n-2} \circ A_{n-1} \circ A_n \circ A_1 \tag{2.114}$$

$$= A_2 \circ A_3 \circ A_4 \circ A_5 \circ \cdots \circ A_{n-1} \circ A_n \circ A_1 \circ A_2 \tag{2.115}$$
$$\vdots$$

$$= A_n \circ A_1 \circ A_2 \circ A_3 \circ \cdots \circ A_{n-3} \circ A_{n-2} \circ A_{n-1} \circ A_n. \tag{2.116}$$

Example 2.11. In the lowest non-binary case $n = 3$, we have the ternary braid operator $C_{\mathcal{V}^3}$: $\mathcal{V} \otimes \mathcal{V} \otimes \mathcal{V} \to \mathcal{V} \otimes \mathcal{V} \otimes \mathcal{V}$ and two ternary braid equations on $\mathcal{V}^{\otimes 5}$

$$\begin{aligned}
&(C_{\mathcal{V}^3} \otimes \mathrm{id}_{\mathcal{V}} \otimes \mathrm{id}_{\mathcal{V}}) \circ (\mathrm{id}_{\mathcal{V}} \otimes C_{\mathcal{V}^3} \otimes \mathrm{id}_{\mathcal{V}}) \circ (\mathrm{id}_{\mathcal{V}} \otimes \mathrm{id}_{\mathcal{V}} \otimes C_{\mathcal{V}^3}) \circ (C_{\mathcal{V}^3} \otimes \mathrm{id}_{\mathcal{V}} \otimes \mathrm{id}_{\mathcal{V}}) \\
&= (\mathrm{id}_{\mathcal{V}} \otimes C_{\mathcal{V}^3} \otimes \mathrm{id}_{\mathcal{V}}) \circ (\mathrm{id}_{\mathcal{V}} \otimes \mathrm{id}_{\mathcal{V}} \otimes C_{\mathcal{V}^3}) \circ (C_{\mathcal{V}^3} \otimes \mathrm{id}_{\mathcal{V}} \otimes \mathrm{id}_{\mathcal{V}}) \circ (\mathrm{id}_{\mathcal{V}} \otimes C_{\mathcal{V}^3} \otimes \mathrm{id}_{\mathcal{V}}) \\
&= (\mathrm{id}_{\mathcal{V}} \otimes \mathrm{id}_{\mathcal{V}} \otimes C_{\mathcal{V}^3}) \circ (C_{\mathcal{V}^3} \otimes \mathrm{id}_{\mathcal{V}} \otimes \mathrm{id}_{\mathcal{V}}) \circ (\mathrm{id}_{\mathcal{V}} \otimes C_{\mathcal{V}^3} \otimes \mathrm{id}_{\mathcal{V}}) \circ (\mathrm{id}_{\mathcal{V}} \otimes \mathrm{id}_{\mathcal{V}} \otimes C_{\mathcal{V}^3}).
\end{aligned} \tag{2.117}$$

Note that the higher braid equations presented above differ from the generalized Yang–Baxter equations of Rowell *et al* (2010), Kitaev and Wang (2012), and Chen (2012a).

The higher braid operators (2.111) satisfying the higher braid equations (2.114)–(2.116) can represent the higher braid group (Duplij 2021a) using (2.6) and (2.113). By analogy with (2.6), we introduce m operators by

$$\mathrm{B}_i(m): \overbrace{\mathcal{V} \otimes \cdots \otimes \mathcal{V}}^{m+n-2} \to \overbrace{\mathcal{V} \otimes \cdots \otimes \mathcal{V}}^{m+n-2}, \quad \mathrm{B}_0(m) = (\mathrm{id}_{\mathcal{V}})^{\otimes(m+n-2)}, \tag{2.118}$$

$$\mathrm{B}_i(m) = \mathrm{id}_{\mathcal{V}}^{\otimes(i-1)} \otimes C_{\mathcal{V}^n} \otimes \mathrm{id}_{\mathcal{V}}^{\otimes(m-i-1)}, \quad i = 1, \ldots, m-1. \tag{2.119}$$

The representation $\pi_m^{[n]}$ of the higher braid group $\mathscr{B}_m^{[n+1]}$ (of $(n+1)$-degree in the notation of Duplij 2021a, 2022) (having $m-1$ generators $\boldsymbol{\sigma}_i$ and identity $\mathbf{e}$) is given by

$$\pi_m^{[n]}: \mathscr{B}_m^{[n+1]} \longrightarrow \mathrm{End}\, \mathcal{V}^{\otimes(m+n-2)}, \tag{2.120}$$

$$\pi_m^{[n]}(\boldsymbol{\sigma}_i) = \mathrm{B}_i(m), \quad i = 1, \ldots, m-1. \tag{2.121}$$

In this way, the generators $\boldsymbol{\sigma}_i$ of the higher braid group $\mathscr{B}_m^{[n+1]}$ satisfy the relations

- *n higher braid relations*

$$\overbrace{\boldsymbol{\sigma}_i \boldsymbol{\sigma}_{i+1} \cdots \boldsymbol{\sigma}_{i+n-2} \boldsymbol{\sigma}_{i+n-1} \boldsymbol{\sigma}_i}^{n+1} \tag{2.122}$$

$$= \boldsymbol{\sigma}_{i+1}\boldsymbol{\sigma}_{i+2}\cdots\boldsymbol{\sigma}_{i+n-1}\boldsymbol{\sigma}_i\boldsymbol{\sigma}_{i+1} \tag{2.123}$$

$$\vdots$$

$$= \boldsymbol{\sigma}_{i+n-1}\boldsymbol{\sigma}_i\boldsymbol{\sigma}_{i+1}\boldsymbol{\sigma}_{i+2}\cdots\boldsymbol{\sigma}_i\boldsymbol{\sigma}_{i+1}\boldsymbol{\sigma}_{i+n-1}, \tag{2.124}$$

$$i = 1, \ldots, m - n, \tag{2.125}$$

- *n-ary far commutativity*

$$\overbrace{\boldsymbol{\sigma}_{i_1}\boldsymbol{\sigma}_{i_2}\cdots\boldsymbol{\sigma}_{i_{n-2}}\boldsymbol{\sigma}_{i_{n-1}}\boldsymbol{\sigma}_{i_n}}^{n} \tag{2.126}$$

$$\vdots$$

$$= \boldsymbol{\sigma}_{\tau(i_1)}\boldsymbol{\sigma}_{\tau(i_2)}\cdots\boldsymbol{\sigma}_{\tau(i_{n-2})}\boldsymbol{\sigma}_{\tau(i_{n-1})}\boldsymbol{\sigma}_{\tau(i_n)}, \tag{2.127}$$

$$\text{if all } \left| i_p - i_s \right| \geqslant n, \quad p, s = 1, \ldots, n, \tag{2.128}$$

where τ is an element of the permutation symmetry group $\tau \in S_n$. The relations (2.122)–(2.127) coincide with those from Duplij (2021a, 2022), obtained by another method, i.e., via the polyadic-binary correspondence.

In the case $m = 4$ and $n = 3$, the higher braid group $\mathscr{B}_4^{[4]}$ is represented by (2.117) and generated by three generators $\boldsymbol{\sigma}_1$, $\boldsymbol{\sigma}_2$, $\boldsymbol{\sigma}_3$, which satisfy two braid relations only (without far commutativity)

$$\boldsymbol{\sigma}_1\boldsymbol{\sigma}_2\boldsymbol{\sigma}_3\boldsymbol{\sigma}_1 = \boldsymbol{\sigma}_2\boldsymbol{\sigma}_3\boldsymbol{\sigma}_1\boldsymbol{\sigma}_2 = \boldsymbol{\sigma}_3\boldsymbol{\sigma}_1\boldsymbol{\sigma}_2\boldsymbol{\sigma}_3. \tag{2.129}$$

According to (2.126)–(2.127), the far commutativity relations appear when the number of elements of the higher braid groups satisfy

$$m \geqslant m_{\min} = n(n - 1) + 2, \tag{2.130}$$

such that all conditions (2.128) should hold. Thus, to have the far commutativity relations in the ordinary (binary) braid group (2.5), we need three generators and $\mathscr{B}_4$, while for $n = 3$ we need at least seven generators $\boldsymbol{\sigma}_i$ and $\mathscr{B}_8^{[4]}$ (see *example* **7.12** in Duplij 2021a).

In the concrete realization of $\mathcal{V}$ as a d-dimensional Euclidean vector space V_d over the complex numbers $\mathbb{C}$ and basis $\{e_i\}$, $i = 1, \ldots, d$, the polyadic (n-ary) braid operator $C_{\mathcal{V}^n}$ becomes a matrix C_{d^n} of size $d^n \times d^n$ which satisfies $n - 1$ higher braid equations (2.114)–(2.116) in matrix form. In the components, the matrix braid operator is

$$C_{d^n} \circ (e_{i_1} \otimes e_{i_2} \otimes \cdots \otimes e_{i_n}) = \sum_{j_1', j_2' \cdots j_n' = 1}^{d} c_{i_1 i_2 \cdots i_n}^{j_1' j_2' \cdots j_n'} \cdot e_{j_1'} \otimes e_{j_2'} \otimes \cdots \otimes e_{j_n'}. \tag{2.131}$$

Thus, we have d^{2n} entries (unknowns) in C_{d^n} satisfying $(n - 1)d^{4n-2}$ equations (2.114)–(2.116) in components of polynomial power $n + 1$. In the minimal non-binary case $n = 3$, we have $2d^{10}$ equations of power 4 for d^6 unknowns, e.g., even for $d = 2$, we have 2048 for 64 components, and for $d = 3$ there are 118 098 equations for

729 components. Thus, solving the matrix higher braid equations directly is cumbersome and only particular cases are possible to investigate, for instance by using permutation matrices (2.28), or the star and circle matrices (2.77)–(2.78).

2.3 Solutions to the ternary braid equations

Here we consider some special solutions to the minimal ternary version ($n = 3$) of the polyadic braid equation (2.114)–(2.116), the ternary braid equation (2.117).

2.3.1 Constant matrix solutions

Let us consider the following two-dimensional vector space $V \equiv V_{d=2}$ (which is important for quantum computations) and the component matrix realization (2.131) of the ternary braiding operator C_8: $V \otimes V \otimes V \to V \otimes V \otimes V$ as

$$C_8 \circ (e_{i_1} \otimes e_{i_2} \otimes e_{i_3}) = \sum_{j_1', j_2', j_3' = 1}^{2} c_{i_1 i_2 i_3}^{j_1' j_2' j_3'} \cdot e_{j_1'} \otimes e_{j_2'} \otimes e_{j_3'}, \quad i_{1,2,3}, j'_{1,2,3} = 1, 2 \tag{2.132}$$

We now turn (2.132) to the standard matrix form (just to fix notations) by introducing the 8-dimensional vector space $\tilde{V}_8 = V \otimes V \otimes V$ with the natural basis $\tilde{e}_{\tilde{k}} = \{e_1 \otimes e_1 \otimes e_1, e_1 \otimes e_1 \otimes e_2, \ldots, e_2 \otimes e_2 \otimes e_2\}$, where $\tilde{k} = 1, \ldots, 8$ is a cumulative index. The linear operator $\tilde{C}_8$: $\tilde{V}_8 \to \tilde{V}_8$ corresponding to (2.132) is given by the 8×8 matrix $\tilde{c}_{\tilde{i}\tilde{j}}$ as $\tilde{C}_8 \circ \tilde{e}_{\tilde{i}} = \sum_{\tilde{j}=1}^{8} \tilde{c}_{\tilde{i}\tilde{j}} \cdot \tilde{e}_{\tilde{j}}$. The operators (2.112)–(2.113) become three 32×32 matrices $\tilde{A}_{1,2,3}$ as

$$\tilde{A}_1 = \tilde{c} \otimes_{\mathrm{K}} I_2 \otimes_{\mathrm{K}} I_2, \quad \tilde{A}_2 = I_2 \otimes_{\mathrm{K}} \tilde{c} \otimes_{\mathrm{K}} I_2, \quad \tilde{A}_3 = I_2 \otimes_{\mathrm{K}} I_2 \otimes_{\mathrm{K}} \tilde{c}, \tag{2.133}$$

where $\otimes_{\mathrm{K}}$ is the Kronecker product of matrices and I_2 is the 2×2 identity matrix. In this notation, the operator ternary braid equations (2.117) become the matrix equations (cf (2.114)–(2.116)) with $n = 3$

$$\tilde{A}_1 \tilde{A}_2 \tilde{A}_3 \tilde{A}_1 = \tilde{A}_2 \tilde{A}_3 \tilde{A}_1 \tilde{A}_2 = \tilde{A}_3 \tilde{A}_1 \tilde{A}_2 \tilde{A}_3, \tag{2.134}$$

which we call the *total matrix ternary braid equations*. Some weaker versions of ternary braiding are described by the *partial braid equations*

$$\text{partial 12–braid equation } \tilde{A}_1 \tilde{A}_2 \tilde{A}_3 \tilde{A}_1 = \tilde{A}_2 \tilde{A}_3 \tilde{A}_1 \tilde{A}_2, \tag{2.135}$$

$$\text{partial 13–braid equation } \tilde{A}_1 \tilde{A}_2 \tilde{A}_3 \tilde{A}_1 = \tilde{A}_3 \tilde{A}_1 \tilde{A}_2 \tilde{A}_3, \tag{2.136}$$

$$\text{partial 23–braid equation } \tilde{A}_2 \tilde{A}_3 \tilde{A}_1 \tilde{A}_2 = \tilde{A}_3 \tilde{A}_1 \tilde{A}_2 \tilde{A}_3, \tag{2.137}$$

where, obviously, two of them are independent. It follows from (2.114)–(2.116) that the weaker versions of braiding are possible for $n \geqslant 3$, but only for higher than binary braiding (the Yang–Baxter equation (2.8)).

Thus, by comparing (2.134) and (2.129) we conclude that (for each invertible matrix $\tilde{c}$ in (2.133) satisfying (2.134)) the isomorphism $\tilde{\pi}_4^{[4]}$: $\sigma_i \mapsto \tilde{A}_i$, $i = 1, 2, 3$ gives a representation of the braid group $\mathcal{B}_4^{[4]}$ by 32×32 matrices over $\mathbb{C}$.

Now we can generate families of solutions corresponding to (2.133)–(2.134) in the following way. Consider an invertible operator Q: $V \to V$ in the two-dimensional vector space $V \equiv V_{d=2}$. In the basis $\{e_1, e_2\}$, its 2×2 matrix q is given by $Q \circ e_i = \sum_{j=1}^{2} q_{ij} \cdot e_j$. In the natural 8-dimensional basis $\tilde{e}_{\tilde{k}}$, the tensor product of operators $Q \otimes Q \otimes Q$ is presented by the Kronecker product of matrices $\tilde{q}_8 = q \otimes_K q \otimes_K q$. Let the 8×8 matrix $\tilde{c}$ be a fixed solution to the ternary braid matrix equations (2.134). Then, the family of solutions $\tilde{c}(q)$ corresponding to the invertible 2×2 matrix q is the conjugation of $\tilde{c}$ by $\tilde{q}_8$ so that

$$\tilde{c}(q) = \tilde{q}_8 \tilde{c} \tilde{q}_8^{-1} = (q \otimes_K q \otimes_K q)\tilde{c}(q^{-1} \otimes_K q^{-1} \otimes_K q^{-1}). \tag{2.138}$$

This also follows directly from the conjugation of the braid equations (2.134)–(2.137) by $q \otimes_K q \otimes_K q \otimes_K q \otimes_K q$ and (2.133). If we include the obvious invariance of the braid equations with the respect of an overall factor $t \in \mathbb{C}$, then the general family of solutions becomes (cf the Yang–Baxter equation Hietarinta 1993)

$$\tilde{c}(q, t) = t\tilde{q}_8 \tilde{c} \tilde{q}_8^{-1} = t(q \otimes_K q \otimes_K q)\tilde{c}(q^{-1} \otimes_K q^{-1} \otimes_K q^{-1}). \tag{2.139}$$

Let

$$q = \begin{pmatrix} a & b \\ c & d \end{pmatrix} \in \mathrm{GL}(2, \mathbb{C}), \tag{2.140}$$

and then the manifest form of $\tilde{q}_8$ is

$$\tilde{q}_8 = \begin{pmatrix} a^3 & a^2b & a^2b & ab^2 & a^2b & ab^2 & ab^2 & b^3 \\ a^2c & a^2d & abc & abd & abc & abd & b^2c & b^2d \\ a^2c & abc & a^2d & abd & abc & b^2c & abd & b^2d \\ ac^2 & acd & acd & ad^2 & bc^2 & bcd & bcd & bd^2 \\ a^2c & abc & abc & b^2c & a^2d & abd & abd & b^2d \\ ac^2 & acd & bc^2 & bcd & acd & ad^2 & bcd & bd^2 \\ ac^2 & bc^2 & acd & bcd & acd & bcd & ad^2 & bd^2 \\ c^3 & c^2d & c^2d & cd^2 & c^2d & cd^2 & cd^2 & d^3 \end{pmatrix}. \tag{2.141}$$

It is important that not every conjugation matrix has this very special form (2.141), and that therefore, in general, conjugated matrices are different solutions of the ternary braid equations (2.134). The matrix $\tilde{q}_8^{\star}\tilde{q}_8$ ($\star$ being the Hermitian conjugation) is diagonal (this case is important for further classification similar to the binary one Dye 2003), when the conditions

$$ab^* + cd^* = 0 \tag{2.142}$$

hold, and so the matrix q has the special form (depending on three complex parameters, for $d \neq 0$)

$$q = \begin{pmatrix} a & b \\ -a\dfrac{b*}{d*} & d \end{pmatrix}. \tag{2.143}$$

We can present the families (2.138) for different ranks because the conjugation by an invertible matrix does not change rank. To avoid demanding (2.142), due to the cumbersome calculations involved, we restrict ourselves to a triangle matrix for q (2.140).

In general, there are $8 \times 8 = 64$ unknowns (elements of the matrix $\tilde{c}$), and each partial braid equation (2.135)–(2.137) gives $32 \times 32 = 1024$ conditions (of power 4) for the elements of $\tilde{c}$, while the total braid equations (2.134) give twice as many conditions $1024 \times 2 = 2048$ (cf the binary case: 64 cubic equations for 16 unknowns (2.8)). This means that, even in the ternary case, the higher braid system of equations is hugely overdetermined and finding even the simplest solutions is a non-trivial task.

2.3.2 Permutation and parameter-permutation 8-vertex solutions

First we consider the case when $\tilde{c}$ is a binary (or logical) matrix consisting of $\{0, 1\}$ only, and, moreover, it is a permutation matrix (see subsection 2.1.4). In the latter case, $\tilde{c}$ can be considered as a matrix over the field $\mathbb{F}_2$ (Galois field $GF(2)$). In total, there are $8! = 40\,320$ permutation matrices of the size 8×8. All of them are invertible of full rank 8 because they are obtained from the identity matrix by permutation of rows and columns.

We have found the following four invertible 8-vertex permutation matrix solutions to the ternary braid equations (2.134)

$$\tilde{c}^{\,\text{bisymm1}}_{\text{rank}=8} = \begin{pmatrix} 1&0&0&0&0&0&0&0 \\ 0&0&0&0&0&0&1&0 \\ 0&0&1&0&0&0&0&0 \\ 0&0&0&0&1&0&0&0 \\ 0&0&0&1&0&0&0&0 \\ 0&0&0&0&0&1&0&0 \\ 0&1&0&0&0&0&0&0 \\ 0&0&0&0&0&0&0&1 \end{pmatrix},$$

$$\tilde{c}^{\,\text{bisymm2}}_{\text{rank}=8} = \begin{pmatrix} 0&0&0&0&0&0&0&1 \\ 0&1&0&0&0&0&0&0 \\ 0&0&0&0&0&1&0&0 \\ 0&0&0&1&0&0&0&0 \\ 0&0&0&0&1&0&0&0 \\ 0&0&1&0&0&0&0&0 \\ 0&0&0&0&0&0&1&0 \\ 1&0&0&0&0&0&0&0 \end{pmatrix}, \qquad \begin{gathered} \operatorname{tr}\tilde{c} = 4, \\ \det\tilde{c} = 1, \\ \text{eigenvalues:}\{1\}^{[4]},\ \{-1\}^{[4]}, \end{gathered} \tag{2.144}$$

$$\tilde{c}_{\text{rank}=8}^{\text{symm1}} = \begin{pmatrix} 1&0&0&0&0&0&0&0 \\ 0&0&0&0&1&0&0&0 \\ 0&0&0&0&0&0&0&1 \\ 0&0&0&1&0&0&0&0 \\ 0&1&0&0&0&0&0&0 \\ 0&0&0&0&0&1&0&0 \\ 0&0&0&0&0&0&1&0 \\ 0&0&1&0&0&0&0&0 \end{pmatrix},$$

$$\tilde{c}_{\text{rank}=8}^{\text{symm2}} = \begin{pmatrix} 0&0&0&0&0&1&0&0 \\ 0&1&0&0&0&0&0&0 \\ 0&0&1&0&0&0&0&0 \\ 0&0&0&0&0&0&1&0 \\ 0&0&0&0&1&0&0&0 \\ 1&0&0&0&0&0&0&0 \\ 0&0&0&1&0&0&0&0 \\ 0&0&0&0&0&0&0&1 \end{pmatrix}, \quad \begin{array}{c} \operatorname{tr}\tilde{c} = 4, \\ \det\tilde{c} = 1, \\ \text{eigenvalues:}\{1\}^{[4]},\ \{-1\}^{[4]}. \end{array} \tag{2.145}$$

The first two solutions (2.144) are given by bisymmetric permutation matrices (see (2.31)), and we call them 8-vertex bisymm1 and bisymm2, respectively. The second two solutions (2.145) are symmetric matrices only (we call them 8-vertex symm1 and symm2), but one matrix is a reflection of the other with respect to the minor diagonal (making them mutually persymmetric). No 90°-symmetric (see (2.32)) solution for the ternary braid equations (2.134) was found. The bisymmetric and symmetric matrices have the same eigenvalues, and are therefore pairwise conjugate but not q-conjugate because the conjugation matrices do not have the form (2.141). Thus, they are four different permutation solutions to the ternary braid equations (2.134). Note that the bisymm1 solution (2.144) coincides with the three-qubit swap operator introduced in Ballard and Wu (2011b).

All the permutation solutions are reflections (or involutions) $\tilde{c}^2 = I_8$ having $\det\tilde{c} = +1$, eigenvalues $\{1, -1\}$, and are semi-magic squares (the sums in rows and columns are 1, but not the sums in both diagonals). The 8-vertex permutation matrix solutions do not form a binary or ternary group because they are not closed with respect to multiplication.

By analogy with (2.33)–(2.34), we obtain the 8-vertex parameter-permutation solutions from (2.144)–(2.145) by replacing units with parameters and then solving the ternary braid equations (2.134). Each type of the permutation solutions bisymm1, 2 and symm1, 2 from (2.144)–(2.145) will give a corresponding series of parameter-permutation solutions over $\mathbb{C}$. The ternary braid maps are determined up to a general complex factor (see (2.14) for the Yang–Baxter maps and (2.139)), and therefore we can present all the parameter-permutation solutions in polynomial form.

- The bisymm1 series consists of two two-parameter matrices with two two-parameter matrices

$$\tilde{c}_{\mathrm{rank}=8}^{\mathrm{bisymm1,\,1}}(x,\, y) = \begin{pmatrix} xy & 0 & 0 & 0 & 0 & 0 & 0 & 0 \\ 0 & 0 & 0 & 0 & 0 & 0 & \pm y^2 & 0 \\ 0 & 0 & xy & 0 & 0 & 0 & 0 & 0 \\ 0 & 0 & 0 & 0 & \pm x^2 & 0 & 0 & 0 \\ 0 & 0 & 0 & \pm y^2 & 0 & 0 & 0 & 0 \\ 0 & 0 & 0 & 0 & 0 & xy & 0 & 0 \\ 0 & \pm x^2 & 0 & 0 & 0 & 0 & 0 & 0 \\ 0 & 0 & 0 & 0 & 0 & 0 & 0 & xy \end{pmatrix}, \tag{2.146}$$

$$\mathrm{tr}\,\tilde{c} = 4xy,$$

$$\det \tilde{c} = x^8 y^8, \;\; x,\, y \neq 0,$$

$$\text{eigenvalues: } \{xy\}^{[6]},\ \{-xy\}^{[2]},$$

$$\tilde{c}_{\mathrm{rank}=8}^{\mathrm{bisymm1,\,2}}(x,\, y) = \begin{pmatrix} xy & 0 & 0 & 0 & 0 & 0 & 0 & 0 \\ 0 & 0 & 0 & 0 & 0 & 0 & \pm y^2 & 0 \\ 0 & 0 & xy & 0 & 0 & 0 & 0 & 0 \\ 0 & 0 & 0 & 0 & \pm x^2 & 0 & 0 & 0 \\ 0 & 0 & 0 & \mp y^2 & 0 & 0 & 0 & 0 \\ 0 & 0 & 0 & 0 & 0 & xy & 0 & 0 \\ 0 & \mp x^2 & 0 & 0 & 0 & 0 & 0 & 0 \\ 0 & 0 & 0 & 0 & 0 & 0 & 0 & xy \end{pmatrix}, \tag{2.147}$$

$$\mathrm{tr}\,\tilde{c} = 4xy,$$

$$\det \tilde{c} = x^8 y^8, \;\; x,\, y \neq 0,$$

$$\text{eigenvalues: } \{xy\}^{[4]},\ \{ixy\}^{[2]},\ \{-ixy\}^{[2]}.$$

- The bisymm2 series consists of four two-parameter matrices

$$\tilde{c}_{\mathrm{rank}=8}^{\mathrm{bisymm2,\,1}}(x,\, y) = \begin{pmatrix} 0 & 0 & 0 & 0 & 0 & 0 & 0 & x^6 \\ 0 & \pm x^3y^3 & 0 & 0 & 0 & 0 & 0 & 0 \\ 0 & 0 & 0 & 0 & 0 & x^4y^2 & 0 & 0 \\ 0 & 0 & 0 & \pm x^3y^3 & 0 & 0 & 0 & 0 \\ 0 & 0 & 0 & 0 & \pm x^3y^3 & 0 & 0 & 0 \\ 0 & 0 & x^2y^4 & 0 & 0 & 0 & 0 & 0 \\ 0 & 0 & 0 & 0 & 0 & 0 & \pm x^3y^3 & 0 \\ y^6 & 0 & 0 & 0 & 0 & 0 & 0 & 0 \end{pmatrix}, \tag{2.148}$$

$$\mathrm{tr}\,\tilde{c} = \pm 4x^3y^3,$$

$$\det \tilde{c}_{\mathrm{rank}=8}^{\mathrm{bisymm2}}(x,\, y) = x^{24}y^{24},\ x,\, y \neq 0,$$

$$\text{eigenvalues : } \{x^3y^3\}^{[2]},\ \{-x^3y^3\}^{[2]},\ \{\pm x^3y^3\}^{[4]},$$

$$\tilde{c}^{\text{bisymm2},\,2}_{\text{rank}=8}(x,\,y)=\begin{pmatrix} 0 & 0 & 0 & 0 & 0 & 0 & 0 & x^6 \\ 0 & \pm x^3y^3 & 0 & 0 & 0 & 0 & 0 & 0 \\ 0 & 0 & 0 & 0 & 0 & x^4y^2 & 0 & 0 \\ 0 & 0 & 0 & \pm x^3y^3 & 0 & 0 & 0 & 0 \\ 0 & 0 & 0 & 0 & \pm x^3y^3 & 0 & 0 & 0 \\ 0 & 0 & -x^2y^4 & 0 & 0 & 0 & 0 & 0 \\ 0 & 0 & 0 & 0 & 0 & 0 & \pm x^3y^3 & 0 \\ -y^6 & 0 & 0 & 0 & 0 & 0 & 0 & 0 \end{pmatrix}, \tag{2.149}$$

$$\text{tr}\,\tilde{c} = \pm 4x^3y^3$$

$$\det \tilde{c}^{\text{bisymm2}}_{\text{rank}=8}(x,\,y) = x^{24}y^{24},\ x,\,y \neq 0,$$

$$\text{eigenvalues}:\ \{ix^3y^3\}^{[2]},\ \{-ix^3y^3\}^{[2]},\ \{\pm x^3y^3\}^{[4]}.$$

- The symm1 series consists of four two-parameter matrices

$$\tilde{c}^{\text{symm1},\,1}_{\text{rank}=8}(x,\,y)=\begin{pmatrix} xy & 0 & 0 & 0 & 0 & 0 & 0 & 0 \\ 0 & 0 & 0 & 0 & \pm xy & 0 & 0 & 0 \\ 0 & 0 & 0 & 0 & 0 & 0 & 0 & y^2 \\ 0 & 0 & 0 & xy & 0 & 0 & 0 & 0 \\ 0 & \pm xy & 0 & 0 & 0 & 0 & 0 & 0 \\ 0 & 0 & 0 & 0 & 0 & xy & 0 & 0 \\ 0 & 0 & 0 & 0 & 0 & 0 & xy & 0 \\ 0 & 0 & x^2 & 0 & 0 & 0 & 0 & 0 \end{pmatrix}, \tag{2.150}$$

$$\text{tr}\,\tilde{c} = 4xy,$$

$$\det \tilde{c} = x^8y^8,\quad x,\,y \neq 0,$$

$$\text{eigenvalues: } \{xy\}^{[6]},\ \{-xy\}^{[2]},$$

$$\tilde{c}^{\text{symm1},\,2}_{\text{rank}=8}(x,\,y)=\begin{pmatrix} xy & 0 & 0 & 0 & 0 & 0 & 0 & 0 \\ 0 & 0 & 0 & 0 & \pm xy & 0 & 0 & 0 \\ 0 & 0 & 0 & 0 & 0 & 0 & 0 & y^2 \\ 0 & 0 & 0 & xy & 0 & 0 & 0 & 0 \\ 0 & \mp xy & 0 & 0 & 0 & 0 & 0 & 0 \\ 0 & 0 & 0 & 0 & 0 & xy & 0 & 0 \\ 0 & 0 & 0 & 0 & 0 & 0 & xy & 0 \\ 0 & 0 & -x^2 & 0 & 0 & 0 & 0 & 0 \end{pmatrix}, \tag{2.151}$$

$$\text{tr}\,\tilde{c} = 4xy,$$

$$\det \tilde{c} = x^8y^8,\quad x,\,y \neq 0,$$

$$\text{eigenvalues: } \{xy\}^{[6]},\ \{-xy\}^{[2]},$$

• The symm2 series consists of four two-parameter matrices

$$\tilde{c}_{\text{rank}=8}^{\text{symm2},\,1}(x,y)=\begin{pmatrix} 0 & 0 & 0 & 0 & 0 & y^2 & 0 & 0 \\ 0 & xy & 0 & 0 & 0 & 0 & 0 & 0 \\ 0 & 0 & xy & 0 & 0 & 0 & 0 & 0 \\ 0 & 0 & 0 & 0 & 0 & 0 & \pm xy & 0 \\ 0 & 0 & 0 & 0 & xy & 0 & 0 & 0 \\ x^2 & 0 & 0 & 0 & 0 & 0 & 0 & 0 \\ 0 & 0 & 0 & \pm xy & 0 & 0 & 0 & 0 \\ 0 & 0 & 0 & 0 & 0 & 0 & 0 & xy \end{pmatrix}, \tag{2.152}$$

$$\operatorname{tr}\tilde{c} = 4xy,$$
$$\det\tilde{c} = x^8y^8,\quad x,y\neq 0,$$
$$\text{eigenvalues: } \{xy\}^{[4]},\ \{ixy\}^{[2]},\ \{-ixy\}^{[2]}.$$

$$\tilde{c}_{\text{rank}=8}^{\text{symm2},\,2}(x,y)=\begin{pmatrix} 0 & 0 & 0 & 0 & 0 & y^2 & 0 & 0 \\ 0 & xy & 0 & 0 & 0 & 0 & 0 & 0 \\ 0 & 0 & xy & 0 & 0 & 0 & 0 & 0 \\ 0 & 0 & 0 & 0 & 0 & 0 & \pm xy & 0 \\ 0 & 0 & 0 & 0 & xy & 0 & 0 & 0 \\ -x^2 & 0 & 0 & 0 & 0 & 0 & 0 & 0 \\ 0 & 0 & 0 & \mp xy & 0 & 0 & 0 & 0 \\ 0 & 0 & 0 & 0 & 0 & 0 & 0 & xy \end{pmatrix}, \tag{2.153}$$

$$\operatorname{tr}\tilde{c} = 4xy,$$
$$\det\tilde{c} = x^8y^8,\quad x,y\neq 0,$$
$$\text{eigenvalues: } \{xy\}^{[4]},\ \{ixy\}^{[2]},\ \{-ixy\}^{[2]}.$$

The above matrices with the same eigenvalues are similar but their conjugation matrices do not have the form of the triple Kronecker product (2.141), and therefore all of them together are 16 different two-parameter invertible solutions to the ternary braid equations (2.134). Further families of solutions can be obtained using ternary q-conjugation (2.139).

2.3.3 Group structure of the star and circle 8-vertex matrices

Here we investigate the group structure of 8×8 matrices by analogy with the star-like (2.35) and circle-like (2.36) 4×4 matrices, which are connected with our 8-vertex constant solutions (2.146)–(2.153) to the ternary braid equations (2.134).

Let us introduce the *star-like* 8×8 *matrices* (cf (2.35)), which correspond to the bisymm series (2.146)–(2.149)

$$N'_{\text{star1}} = \begin{pmatrix} x & 0 & 0 & 0 & 0 & 0 & 0 & 0 \\ 0 & 0 & 0 & 0 & 0 & 0 & y & 0 \\ 0 & 0 & z & 0 & 0 & 0 & 0 & 0 \\ 0 & 0 & 0 & 0 & s & 0 & 0 & 0 \\ 0 & 0 & 0 & t & 0 & 0 & 0 & 0 \\ 0 & 0 & 0 & 0 & 0 & u & 0 & 0 \\ 0 & v & 0 & 0 & 0 & 0 & 0 & 0 \\ 0 & 0 & 0 & 0 & 0 & 0 & 0 & w \end{pmatrix}, \quad N'_{\text{star2}} = \begin{pmatrix} 0 & 0 & 0 & 0 & 0 & 0 & 0 & y \\ 0 & x & 0 & 0 & 0 & 0 & 0 & 0 \\ 0 & 0 & 0 & 0 & 0 & s & 0 & 0 \\ 0 & 0 & 0 & z & 0 & 0 & 0 & 0 \\ 0 & 0 & 0 & 0 & u & 0 & 0 & 0 \\ 0 & 0 & t & 0 & 0 & 0 & 0 & 0 \\ 0 & 0 & 0 & 0 & 0 & 0 & w & 0 \\ v & 0 & 0 & 0 & 0 & 0 & 0 & 0 \end{pmatrix}, \tag{2.154}$$

$$\operatorname{tr} N' = x + z + u + w, \quad \det N' = stuvwxyz, \quad s, t, u, v, w, x, y, z \neq 0,$$
$$\text{eigenvalues: } x, z, u, w, -\sqrt{yv}, \sqrt{yv}, -\sqrt{st}, \sqrt{st},$$

and the *circle-like* 8×8 *matrices* (cf (2.36)), which correspond to the **symm** series (2.150)–(2.153)

$$N'_{\text{circ1}} = \begin{pmatrix} x & 0 & 0 & 0 & 0 & 0 & 0 & 0 \\ 0 & 0 & 0 & 0 & y & 0 & 0 & 0 \\ 0 & 0 & 0 & 0 & 0 & 0 & 0 & z \\ 0 & 0 & 0 & s & 0 & 0 & 0 & 0 \\ 0 & t & 0 & 0 & 0 & 0 & 0 & 0 \\ 0 & 0 & 0 & 0 & 0 & u & 0 & 0 \\ 0 & 0 & 0 & 0 & 0 & 0 & v & 0 \\ 0 & 0 & w & 0 & 0 & 0 & 0 & 0 \end{pmatrix}, \quad N'_{\text{circ2}} = \begin{pmatrix} 0 & 0 & 0 & 0 & 0 & y & 0 & 0 \\ 0 & x & 0 & 0 & 0 & 0 & 0 & 0 \\ 0 & 0 & s & 0 & 0 & 0 & 0 & 0 \\ 0 & 0 & 0 & 0 & 0 & 0 & z & 0 \\ 0 & 0 & 0 & 0 & u & 0 & 0 & 0 \\ t & 0 & 0 & 0 & 0 & 0 & 0 & 0 \\ 0 & 0 & 0 & w & 0 & 0 & 0 & 0 \\ 0 & 0 & 0 & 0 & 0 & 0 & 0 & v \end{pmatrix}, \tag{2.155}$$

$$\begin{gathered} \operatorname{tr} N' = x + s + u + v, \quad \det N' = stuvwxyz, \quad s, t, u, v, w, x, y, z \neq 0, \\ \text{eigenvalues: } x, s, u, v, -\sqrt{ty}, \sqrt{ty}, -\sqrt{wz}, \sqrt{wz}. \end{gathered} \tag{2.156}$$

We denote the corresponding sets by $\mathsf{N}'_{\text{star1, 2}} = \left\{N'_{\text{star1, 2}}\right\}$ and $\mathsf{N}'_{\text{circ1, 2}} = \left\{N'_{\text{circ1, 2}}\right\}$, and then we have for them (which differs from 4×4 matrix sets (2.41))

$$\mathsf{M}'_{\text{full}} = \mathsf{N}'_{\text{star1}} \bigcup \mathsf{N}'_{\text{star2}} \bigcup \mathsf{N}'_{\text{circ1}} \bigcup \mathsf{N}'_{\text{circ2}}, \quad \mathsf{N}'_{\text{star1}} \bigcap \mathsf{N}'_{\text{star2}} \bigcap \mathsf{N}'_{\text{circ1}} \bigcap \mathsf{N}'_{\text{circ2}} = \mathsf{D}, \tag{2.157}$$

where D is the set of diagonal 8×8 matrices. Again, as for 4×4 star-like and circle-like matrices, there are no closed binary multiplications among the sets of 8-vertex matrices (2.154)–(2.155). Nevertheless, we have the following triple set products

$$\mathsf{N}'_{\text{star1}} \mathsf{N}'_{\text{star1}} \mathsf{N}'_{\text{star1}} = \mathsf{N}'_{\text{star1}}, \tag{2.158}$$

$$\mathsf{N}'_{\text{star2}} \mathsf{N}'_{\text{star2}} \mathsf{N}'_{\text{star2}} = \mathsf{N}'_{\text{star2}}, \tag{2.159}$$

$$\mathsf{N}'_{\text{circ1}} \mathsf{N}'_{\text{circ1}} \mathsf{N}'_{\text{circ1}} = \mathsf{N}'_{\text{circ1}}, \tag{2.160}$$

$$\mathsf{N}'_{\text{circ2}} \mathsf{N}'_{\text{circ2}} \mathsf{N}'_{\text{circ2}} = \mathsf{N}'_{\text{circ2}}, \tag{2.161}$$

which should be compared with the analogous 4×4 matrices (2.39)–(2.40): note that now we do not have pentuple products.

Using the definitions (2.42)–(2.45), we interpret the closed products (2.158)–(2.159) and (2.160)–(2.161) as the multiplications $\mu^{[3]}$ (being the ordinary triple matrix product) of the *ternary semigroups* $\mathcal{S}^{[3]}_{\mathrm{star1},\,2}(8,\,\mathbb{C}) = \left\{\mathsf{N}'_{\mathrm{star1},\,2}\middle|\mu^{[3]}\right\}$ and $\mathcal{S}^{[3]}_{\mathrm{circ1},\,2}(8,\,\mathbb{C}) = \left\{\mathsf{N}'_{\mathrm{circ1},\,2}\middle|\mu^{[3]}\right\}$, respectively. The corresponding querelements (2.42) are given by

$$\bar{N}'_{\mathrm{star1}} = N'^{-1}_{\mathrm{star1}} = \begin{pmatrix} \frac{1}{x} & 0 & 0 & 0 & 0 & 0 & 0 & 0 \\ 0 & 0 & 0 & 0 & 0 & 0 & \frac{1}{v} & 0 \\ 0 & 0 & \frac{1}{z} & 0 & 0 & 0 & 0 & 0 \\ 0 & 0 & 0 & 0 & \frac{1}{t} & 0 & 0 & 0 \\ 0 & 0 & 0 & \frac{1}{s} & 0 & 0 & 0 & 0 \\ 0 & 0 & 0 & 0 & 0 & \frac{1}{u} & 0 & 0 \\ 0 & \frac{1}{y} & 0 & 0 & 0 & 0 & 0 & 0 \\ 0 & 0 & 0 & 0 & 0 & 0 & 0 & \frac{1}{w} \end{pmatrix}, \tag{2.162}$$

$$\bar{N}'_{\mathrm{star2}} = N'^{-1}_{\mathrm{star2}} = \begin{pmatrix} 0 & 0 & 0 & 0 & 0 & 0 & 0 & \frac{1}{v} \\ 0 & \frac{1}{x} & 0 & 0 & 0 & 0 & 0 & 0 \\ 0 & 0 & 0 & 0 & 0 & \frac{1}{t} & 0 & 0 \\ 0 & 0 & 0 & \frac{1}{z} & 0 & 0 & 0 & 0 \\ 0 & 0 & 0 & 0 & \frac{1}{u} & 0 & 0 & 0 \\ 0 & 0 & \frac{1}{s} & 0 & 0 & 0 & 0 & 0 \\ 0 & 0 & 0 & 0 & 0 & 0 & \frac{1}{w} & 0 \\ \frac{1}{y} & 0 & 0 & 0 & 0 & 0 & 0 & 0 \end{pmatrix}, \quad s,\, t,\, u,\, v,\, w,\, x,\, y,\, z \neq 0, \tag{2.163}$$

and

$$\bar{N}'_{\mathrm{circ1}} = N'^{-1}_{\mathrm{circ1}} = \begin{pmatrix} \frac{1}{x} & 0 & 0 & 0 & 0 & 0 & 0 & 0 \\ 0 & 0 & 0 & 0 & \frac{1}{t} & 0 & 0 & 0 \\ 0 & 0 & 0 & 0 & 0 & 0 & 0 & \frac{1}{w} \\ 0 & 0 & 0 & \frac{1}{s} & 0 & 0 & 0 & 0 \\ 0 & \frac{1}{y} & 0 & 0 & 0 & 0 & 0 & 0 \\ 0 & 0 & 0 & 0 & 0 & \frac{1}{u} & 0 & 0 \\ 0 & 0 & 0 & 0 & 0 & 0 & \frac{1}{v} & 0 \\ 0 & 0 & \frac{1}{z} & 0 & 0 & 0 & 0 & 0 \end{pmatrix}, \tag{2.164}$$

$$\bar{N}'_{\mathrm{circ2}} = N'^{-1}_{\mathrm{circ2}} = \begin{pmatrix} 0 & 0 & 0 & 0 & 0 & \frac{1}{t} & 0 & 0 \\ 0 & \frac{1}{x} & 0 & 0 & 0 & 0 & 0 & 0 \\ 0 & 0 & \frac{1}{s} & 0 & 0 & 0 & 0 & 0 \\ 0 & 0 & 0 & 0 & 0 & 0 & \frac{1}{w} & 0 \\ 0 & 0 & 0 & 0 & \frac{1}{u} & 0 & 0 & 0 \\ \frac{1}{y} & 0 & 0 & 0 & 0 & 0 & 0 & 0 \\ 0 & 0 & 0 & \frac{1}{z} & 0 & 0 & 0 & 0 \\ 0 & 0 & 0 & 0 & 0 & 0 & 0 & \frac{1}{v} \end{pmatrix}, \quad s, t, u, v, w, x, y, z \neq 0. \tag{2.165}$$

The ternary semigroups $\mathcal{S}^{[3]}_{\mathrm{star1,\,2}}(8, \mathbb{C}) = \left\{ \mathsf{N}'_{\mathrm{star1,\,2}} \middle| \mu^{[3]} \right\}$ and $\mathcal{S}^{[3]}_{\mathrm{circ1,\,2}}(8, \mathbb{C}) = \left\{ \mathsf{N}'_{\mathrm{circ1,\,2}} \middle| \mu^{[3]} \right\}$ in which every element has its querelement given by (2.162)–(2.164) become the *ternary groups* $\mathcal{G}^{[3]}_{\mathrm{star1,\,2}}(8, \mathbb{C}) = \left\{ \mathsf{N}'_{\mathrm{star1,\,2}} \middle| \mu^{[3]}, \overline{(\,)} \right\}$ and $\mathcal{G}^{[3]}_{\mathrm{circ1,\,2}}(8, \mathbb{C}) = \left\{ \mathsf{N}'_{\mathrm{circ1,\,2}} \middle| \mu^{[3]}, \overline{(\,)} \right\}$, which are four different (3-nonderived) ternary subgroups of the derived ternary general linear group $\mathrm{GL}^{[3]}(8, \mathbb{C})$. The ternary identities in $\mathscr{G}^{[3]}_{\text{star1, 2}}(8, \mathbb{C})$ and $\mathscr{G}^{[3]}_{\text{circ1, 2}}(8, \mathbb{C})$ are the following different continuous sets $\mathsf{I}^{'[3]}_{\text{star1, 2}} = \left\{ I^{'[3]}_{\text{star1, 2}} \right\}$ and $\mathsf{I}^{'[3]}_{\text{circ1, 2}} = \left\{ I^{'[3]}_{\text{circ1, 2}} \right\}$, where

$$I^{'[3]}_{\text{star1}} = \begin{pmatrix} e^{i\alpha_1} & 0 & 0 & 0 & 0 & 0 & 0 & 0 \\ 0 & 0 & 0 & 0 & 0 & 0 & e^{i\alpha_2} & 0 \\ 0 & 0 & e^{i\alpha_3} & 0 & 0 & 0 & 0 & 0 \\ 0 & 0 & 0 & 0 & e^{i\alpha_4} & 0 & 0 & 0 \\ 0 & 0 & 0 & e^{i\alpha_5} & 0 & 0 & 0 & 0 \\ 0 & 0 & 0 & 0 & 0 & e^{i\alpha_6} & 0 & 0 \\ 0 & e^{i\alpha_7} & 0 & 0 & 0 & 0 & 0 & 0 \\ 0 & 0 & 0 & 0 & 0 & 0 & 0 & e^{i\alpha_8} \end{pmatrix},$$
$$I^{'[3]}_{\text{star2}} = \begin{pmatrix} 0 & 0 & 0 & 0 & 0 & 0 & 0 & e^{i\alpha_2} \\ 0 & e^{i\alpha_1} & 0 & 0 & 0 & 0 & 0 & 0 \\ 0 & 0 & 0 & 0 & 0 & e^{i\alpha_4} & 0 & 0 \\ 0 & 0 & 0 & e^{i\alpha_3} & 0 & 0 & 0 & 0 \\ 0 & 0 & 0 & 0 & e^{i\alpha_6} & 0 & 0 & 0 \\ 0 & 0 & e^{i\alpha_5} & 0 & 0 & 0 & 0 & 0 \\ 0 & 0 & 0 & 0 & 0 & 0 & e^{i\alpha_8} & 0 \\ e^{i\alpha_7} & 0 & 0 & 0 & 0 & 0 & 0 & 0 \end{pmatrix}, \tag{2.166}$$

$$e^{2i\alpha_1} = e^{2i\alpha_3} = e^{2i\alpha_6} = e^{2i\alpha_8} = e^{i(\alpha_2+\alpha_7)} = e^{i(\alpha_4+\alpha_5)} = 1, \ \ \alpha_1, \ldots, \alpha_8 \in \mathbb{R}, \tag{2.167}$$

and

$$I^{'[3]}_{\text{circ1}} = \begin{pmatrix} e^{i\alpha_1} & 0 & 0 & 0 & 0 & 0 & 0 & 0 \\ 0 & 0 & 0 & 0 & e^{i\alpha_2} & 0 & 0 & 0 \\ 0 & 0 & 0 & 0 & 0 & 0 & 0 & e^{i\alpha_3} \\ 0 & 0 & 0 & e^{i\alpha_4} & 0 & 0 & 0 & 0 \\ 0 & e^{i\alpha_5} & 0 & 0 & 0 & 0 & 0 & 0 \\ 0 & 0 & 0 & 0 & 0 & e^{i\alpha_6} & 0 & 0 \\ 0 & 0 & 0 & 0 & 0 & 0 & e^{i\alpha_7} & 0 \\ 0 & 0 & e^{i\alpha_8} & 0 & 0 & 0 & 0 & 0 \end{pmatrix},$$
$$I^{'[3]}_{\text{circ2}} = \begin{pmatrix} 0 & 0 & 0 & 0 & 0 & e^{i\alpha_2} & 0 & 0 \\ 0 & e^{i\alpha_1} & 0 & 0 & 0 & 0 & 0 & 0 \\ 0 & 0 & e^{i\alpha_4} & 0 & 0 & 0 & 0 & 0 \\ 0 & 0 & 0 & 0 & 0 & 0 & e^{i\alpha_3} & 0 \\ 0 & 0 & 0 & 0 & e^{i\alpha_6} & 0 & 0 & 0 \\ e^{i\alpha_5} & 0 & 0 & 0 & 0 & 0 & 0 & 0 \\ 0 & 0 & 0 & e^{i\alpha_8} & 0 & 0 & 0 & 0 \\ 0 & 0 & 0 & 0 & 0 & 0 & 0 & e^{i\alpha_7} \end{pmatrix}, \tag{2.168}$$

$$e^{2i\alpha_1} = e^{2i\alpha_4} = e^{2i\alpha_6} = e^{2i\alpha_7} = e^{i(\alpha_3+\alpha_8)} = e^{i(\alpha_2+\alpha_5)} = 1, \ \ \alpha_1, \ldots, \alpha_8 \in \mathbb{R}, \tag{2.169}$$

such that all the identities are the 8×8 matrix reflections $(I'^{[3]})^2 = I_8$ (see (2.45)). If $\alpha_j = 0$, $j = 1, \dots, 8$, then the ternary identities (2.167)–(2.169) coincide with the 8×8 permutation matrices (2.144)–(2.145), which are solutions to the ternary braid equations (2.134).

The module structure of the 8-vertex star-like (2.154) and circle-like (2.155) 8×8 matrix sets differs from the 4×4 matrix sets (2.53)–(2.76). First, because of the absence of pentuple matrix products (2.71)–(2.76), and second through some differences in the ternary closed products of sets.

We have the following triple relations between star and circle matrices separately (the sets corresponding to modules are in brackets, and we informally denote modules by their sets)

$$\mathsf{N}'_{\text{star1}}(\mathsf{N}'_{\text{star2}})\mathsf{N}'_{\text{star1}} = (\mathsf{N}'_{\text{star2}}), \quad \mathsf{N}'_{\text{circ1}}(\mathsf{N}'_{\text{circ2}})\mathsf{N}'_{\text{circ1}} = (\mathsf{N}'_{\text{circ2}}), \tag{2.170}$$

$$\mathsf{N}'_{\text{star1}}\mathsf{N}'_{\text{star1}}(\mathsf{N}'_{\text{star2}}) = (\mathsf{N}'_{\text{star2}}), \quad \mathsf{N}'_{\text{circ1}}\mathsf{N}'_{\text{circ1}}(\mathsf{N}'_{\text{circ2}}) = \mathsf{N}'_{\text{circ2}}, \tag{2.171}$$

$$(\mathsf{N}'_{\text{star2}})\mathsf{N}'_{\text{star1}}\mathsf{N}'_{\text{star1}} = (\mathsf{N}'_{\text{star2}}), \quad (\mathsf{N}'_{\text{circ2}})\mathsf{N}'_{\text{circ1}}\mathsf{N}'_{\text{circ1}} = (\mathsf{N}'_{\text{circ2}}), \tag{2.172}$$

$$\mathsf{N}'_{\text{star2}}\mathsf{N}'_{\text{star2}}(\mathsf{N}'_{\text{star1}}) = (\mathsf{N}'_{\text{star1}}), \quad \mathsf{N}'_{\text{circ2}}\mathsf{N}'_{\text{circ2}}(\mathsf{N}'_{\text{circ1}}) = (\mathsf{N}'_{\text{circ1}}), \tag{2.173}$$

$$\mathsf{N}'_{\text{star2}}(\mathsf{N}'_{\text{star1}})\mathsf{N}'_{\text{star2}} = (\mathsf{N}'_{\text{star1}}), \quad \mathsf{N}'_{\text{circ2}}(\mathsf{N}'_{\text{circ1}})\mathsf{N}'_{\text{circ2}} = (\mathsf{N}'_{\text{circ1}}), \tag{2.174}$$

$$(\mathsf{N}'_{\text{star1}})\mathsf{N}'_{\text{star2}}\mathsf{N}'_{\text{star2}} = (\mathsf{N}'_{\text{star1}}), \quad (\mathsf{N}'_{\text{circ1}})\mathsf{N}'_{\text{circ2}}\mathsf{N}'_{\text{circ2}} = (\mathsf{N}'_{\text{circ1}}). \tag{2.175}$$

So we may observe the following module structures: (1) from (2.170)–(2.172), the sets $\mathsf{N}'_{\text{star2}}$ ($\mathsf{N}'_{\text{circ2}}$) are the middle, right, and left ternary modules over $\mathsf{N}'_{\text{star1}}$ ($\mathsf{N}'_{\text{circ1}}$); (2) from (2.173)–(2.175), the set $\mathsf{N}'_{\text{star1}}$ ($\mathsf{N}'_{\text{circ1}}$) are middle, right, and left ternary modules over $\mathsf{N}'_{\text{star2}}$ ($\mathsf{N}'_{\text{circ2}}$);

$$\mathsf{N}'_{\text{star1}}\mathsf{N}'_{\text{star1}}(\mathsf{N}'_{\text{circ1}}) = (\mathsf{N}'_{\text{circ1}}), \quad (\mathsf{N}'_{\text{circ1}})\mathsf{N}'_{\text{star1}}\mathsf{N}'_{\text{star1}} = (\mathsf{N}'_{\text{circ1}}), \tag{2.176}$$

$$\mathsf{N}'_{\text{star1}}\mathsf{N}'_{\text{star1}}(\mathsf{N}'_{\text{circ2}}) = (\mathsf{N}'_{\text{circ2}}), \quad (\mathsf{N}'_{\text{circ2}})\mathsf{N}'_{\text{star1}}\mathsf{N}'_{\text{star1}} = (\mathsf{N}'_{\text{circ2}}), \tag{2.177}$$

$$\mathsf{N}'_{\text{star2}}\mathsf{N}'_{\text{star2}}(\mathsf{N}'_{\text{circ1}}) = (\mathsf{N}'_{\text{circ1}}), \quad (\mathsf{N}'_{\text{circ1}})\mathsf{N}'_{\text{star2}}\mathsf{N}'_{\text{star2}} = (\mathsf{N}'_{\text{circ1}}), \tag{2.178}$$

$$\mathsf{N}'_{\text{star2}}\mathsf{N}'_{\text{star2}}(\mathsf{N}'_{\text{circ2}}) = (\mathsf{N}'_{\text{circ2}}), \quad (\mathsf{N}'_{\text{circ2}})\mathsf{N}'_{\text{star2}}\mathsf{N}'_{\text{star2}} = (\mathsf{N}'_{\text{circ2}}), \tag{2.179}$$

(3) from (2.176)–(2.179), the sets $\mathsf{N}'_{\text{circ1, 2}}$ are right and left ternary modules over $\mathsf{N}'_{\text{star1, 2}}$;

$$\mathsf{N}'_{\text{circ1}}\mathsf{N}'_{\text{circ1}}(\mathsf{N}'_{\text{star1}}) = (\mathsf{N}'_{\text{star1}}), \quad (\mathsf{N}'_{\text{star1}})\mathsf{N}'_{\text{circ1}}\mathsf{N}'_{\text{circ1}} = (\mathsf{N}'_{\text{star1}}), \tag{2.180}$$

$$\mathsf{N}'_{\text{circ1}}\mathsf{N}'_{\text{circ1}}(\mathsf{N}'_{\text{star2}}) = (\mathsf{N}'_{\text{star2}}), \quad (\mathsf{N}'_{\text{star2}})\mathsf{N}'_{\text{circ1}}\mathsf{N}'_{\text{circ1}} = (\mathsf{N}'_{\text{star2}}), \tag{2.181}$$

$$\mathsf{N}'_{\text{circ2}}\mathsf{N}'_{\text{circ2}}(\mathsf{N}'_{\text{star1}}) = (\mathsf{N}'_{\text{star1}}), \quad (\mathsf{N}'_{\text{star1}})\mathsf{N}'_{\text{circ2}}\mathsf{N}'_{\text{circ2}} = (\mathsf{N}'_{\text{star1}}), \tag{2.182}$$

$$\mathsf{N}'_{\text{circ2}}\mathsf{N}'_{\text{circ2}}(\mathsf{N}'_{\text{star2}}) = (\mathsf{N}'_{\text{star2}}), \quad (\mathsf{N}'_{\text{star2}})\mathsf{N}'_{\text{circ2}}\mathsf{N}'_{\text{circ2}} = (\mathsf{N}'_{\text{star2}}), \tag{2.183}$$

(4) from (2.180)–(2.183), the sets $\mathsf{N}'_{\mathrm{star1,\,2}}$ are right and left ternary modules over $\mathsf{N}'_{\mathrm{circ1,\,2}}$.

2.3.4 Group structure of the star and circle 16-vertex matrices

Next we will introduce 8×8 matrices of a special form similar to the star 8-vertex matrices (2.77) and the circle 8-vertex matrices (2.78), analyze their group structure, and establish which ones could be 16-vertex solutions to the ternary braid equations (2.134). We will derive the solutions in the opposite way to that for the 8-vertex Yang–Baxter maps, following the note after (2.34). Indeed, the sum of the permutation bisymm solutions (2.144) gives the shape of the 8×8 star matrix M'_{star} (as in (2.77)), while the sum of symm solutions (2.145) gives the 8×8 circle matrix M'_{circ} (as in (2.78))

$$M'_{\mathrm{star}} = \begin{pmatrix} x & 0 & 0 & 0 & 0 & 0 & 0 & y \\ 0 & z & 0 & 0 & 0 & 0 & s & 0 \\ 0 & 0 & t & 0 & 0 & u & 0 & 0 \\ 0 & 0 & 0 & v & w & 0 & 0 & 0 \\ 0 & 0 & 0 & a & b & 0 & 0 & 0 \\ 0 & 0 & c & 0 & 0 & d & 0 & 0 \\ 0 & f & 0 & 0 & 0 & 0 & g & 0 \\ h & 0 & 0 & 0 & 0 & 0 & 0 & p \end{pmatrix}, \tag{2.184}$$

$$M'_{\mathrm{circ}} = \begin{pmatrix} x & 0 & 0 & 0 & 0 & y & 0 & 0 \\ 0 & z & 0 & 0 & s & 0 & 0 & 0 \\ 0 & 0 & t & 0 & 0 & 0 & 0 & u \\ 0 & 0 & 0 & v & 0 & 0 & w & 0 \\ 0 & f & 0 & 0 & g & 0 & 0 & 0 \\ h & 0 & 0 & 0 & 0 & p & 0 & 0 \\ 0 & 0 & 0 & a & 0 & 0 & b & 0 \\ 0 & 0 & c & 0 & 0 & 0 & 0 & d \end{pmatrix}, \tag{2.185}$$

$$\begin{aligned} \operatorname{tr} M' &= x + z + t + v + b + d + g + p, \\ \det M' &= (bv - aw)(cu - dt)(fs - gz)(px - hy), \end{aligned} \tag{2.186}$$

$$\begin{aligned} &\text{eigenvalues}: \ \frac{1}{2}\left(d + t - \sqrt{4cu + (d-t)^2}\right), \frac{1}{2}\left(d + t + \sqrt{4cu + (d-t)^2}\right), \\ &\frac{1}{2}\left(b + v - \sqrt{4aw + (b-v)^2}\right), \frac{1}{2}\left(b + v + \sqrt{4aw + (b-v)^2}\right), \\ &\frac{1}{2}\left(p + x + \sqrt{4hy + (p-x)^2}\right), \frac{1}{2}\left(p + x - \sqrt{4hy + (p-x)^2}\right), \\ &\frac{1}{2}\left(g + z - \sqrt{4fs + (g-z)^2}\right), \frac{1}{2}\left(g + z + \sqrt{4fs + (g-z)^2}\right). \end{aligned} \tag{2.187}$$

The 16-vertex matrices are invertible, if $\det M'_{\rm star} \neq 0$ and $\det M'_{\rm circ} \neq 0$, which give the following joint conditions on the parameters (cf (2.80))

$$bv - aw \neq 0, \quad cu - dt \neq 0, \quad fs - gz \neq 0, \quad px - hy \neq 0. \tag{2.188}$$

Only in this concrete parametrization (2.184) and (2.185) do the matrices $M'_{\rm star}$ and $M'_{\rm circ}$ have the same trace, determinant, and eigenvalues, and they are diagonalizable and conjugate via

$$U' = \begin{pmatrix} 1 & 0 & 0 & 0 & 0 & 0 & 0 & 0 \\ 0 & 1 & 0 & 0 & 0 & 0 & 0 & 0 \\ 0 & 0 & 1 & 0 & 0 & 0 & 0 & 0 \\ 0 & 0 & 0 & 1 & 0 & 0 & 0 & 0 \\ 0 & 0 & 0 & 0 & 1 & 0 & 1 & 0 \\ 0 & 0 & 0 & 0 & 0 & 0 & 0 & 1 \\ 0 & 0 & 0 & 0 & 1 & 0 & 0 & 0 \\ 0 & 0 & 0 & 0 & 0 & 1 & 0 & 0 \end{pmatrix}. \tag{2.189}$$

The matrix U' cannot be presented in the form of a triple Kronecker product (2.141), and so two matrices $M'_{\rm star}$ and $M'_{\rm circ}$ are not q-conjugate in the parametrization (2.184) and (2.185), and can lead to different solutions to the ternary braid equations (2.134). It follows from (2.188) that 16-vertex matrices with all nonzero entries equal to 1 are noninvertible, having vanishing determinant and rank 4 (despite each one being a sum of two permutation matrices). In the case all the conditions (2.188) holding, the inverse matrices become

$$M'^{-1}_{\rm star} = \begin{pmatrix} \frac{p}{px-hy} & 0 & 0 & 0 & 0 & 0 & 0 & -\frac{y}{px-hy} \\ 0 & -\frac{g}{fs-gz} & 0 & 0 & 0 & 0 & \frac{s}{fs-gz} & 0 \\ 0 & 0 & -\frac{d}{cu-dt} & 0 & 0 & \frac{u}{cu-dt} & 0 & 0 \\ 0 & 0 & 0 & \frac{b}{bv-aw} & -\frac{w}{bv-aw} & 0 & 0 & 0 \\ 0 & 0 & 0 & -\frac{a}{bv-aw} & \frac{v}{bv-aw} & 0 & 0 & 0 \\ 0 & 0 & \frac{c}{cu-dt} & 0 & 0 & -\frac{t}{cu-dt} & 0 & 0 \\ 0 & \frac{f}{fs-gz} & 0 & 0 & 0 & 0 & -\frac{z}{fs-gz} & 0 \\ -\frac{h}{px-hy} & 0 & 0 & 0 & 0 & 0 & 0 & \frac{x}{px-hy} \end{pmatrix}, \tag{2.190}$$

$$M_{\rm circ}^{\prime -1} = \begin{pmatrix} \frac{p}{px-hy} & 0 & 0 & 0 & 0 & -\frac{y}{px-hy} & 0 & 0 \\ 0 & -\frac{g}{fs-gz} & 0 & 0 & \frac{s}{fs-gz} & 0 & 0 & 0 \\ 0 & 0 & -\frac{d}{cu-dt} & 0 & 0 & 0 & 0 & \frac{u}{cu-dt} \\ 0 & 0 & 0 & \frac{b}{bv-aw} & 0 & 0 & -\frac{w}{bv-aw} & 0 \\ 0 & \frac{f}{fs-gz} & 0 & 0 & -\frac{z}{fs-gz} & 0 & 0 & 0 \\ -\frac{h}{px-hy} & 0 & 0 & 0 & 0 & \frac{x}{px-hy} & 0 & 0 \\ 0 & 0 & 0 & -\frac{a}{bv-aw} & 0 & 0 & \frac{v}{bv-aw} & 0 \\ 0 & 0 & \frac{c}{cu-dt} & 0 & 0 & 0 & 0 & -\frac{t}{cu-dt} \end{pmatrix}. \quad (2.191)$$

Denoting the sets of matrices corresponding to (2.184) and (2.185) by $\mathsf{M}'_{\rm star}$ and $\mathsf{M}'_{\rm circ}$, their multiplications are

$$\mathsf{M}'_{\rm star}\mathsf{M}'_{\rm star} = \mathsf{M}'_{\rm star}, \qquad \mathsf{M}'_{\rm circ}\mathsf{M}'_{\rm circ} = \mathsf{M}'_{\rm circ}, \quad (2.192)$$

and in term of sets $\mathsf{M}'_{\rm star} = \mathsf{N}'_{\rm star1} \cup \mathsf{N}'_{\rm star2}$ and $\mathsf{M}'_{\rm circ} = \mathsf{N}'_{\rm circ1} \cup \mathsf{N}'_{\rm circ2}$, and $\mathsf{N}'_{\rm star1} \cap \mathsf{N}'_{\rm star2} = \mathsf{D}$ and $\mathsf{N}'_{\rm circ1} \cap \mathsf{N}'_{\rm circ2} = \mathsf{D}$ (see (2.157)). Note that the structure (2.192) is considerably different from the binary case (2.81)–(2.83), and therefore it may not necessarily be related to the Cartan decomposition.

The products (2.192) mean that both $\mathsf{M}'_{\rm star}$ and $\mathsf{M}'_{\rm circ}$ are separately closed with respect to binary matrix multiplication $(\cdot)$, and therefore $\mathcal{S}^{\rm star}_{\rm 16vert} = \langle \mathsf{M}'_{\rm star} | \cdot \rangle$ and $\mathcal{S}^{\rm circ}_{\rm 16vert} = \langle \mathsf{M}'_{\rm circ} | \cdot \rangle$ are semigroups. We denote their intersection by $\mathcal{S}^{\rm diag}_{8vert} = \mathcal{S}^{\rm star}_{\rm 16vert} \cap \mathcal{S}^{\rm circ}_{\rm 16vert}$ which is a semigroup of diagonal 8-vertex matrices. In case, the invertibility conditions (2.188) are fulfilled, the sets $\mathsf{M}'_{\rm star}$ and $\mathsf{M}'_{\rm circ}$ form subgroups $\mathcal{G}^{\rm star}_{\rm 16vert} = \langle \mathsf{M}'_{\rm star} | \cdot, (\_\_)^{-1}, \mathcal{I}_8 \rangle$ and $\mathcal{G}^{\rm circ}_{\rm 16vert} = \langle \mathsf{M}'_{\rm circ} | \cdot, (\_\_)^{-1}, \mathcal{I}_8 \rangle$ (where I_8 is the 8×8 identity matrix) of $\mathrm{GL}(8, \mathbb{C})$ with the inverse elements given explicitly by (2.190)–(2.191). Because the elements $M'_{\rm star}$ and $M'_{\rm circ}$ in (2.184) and (2.185) are conjugates by the invertible matrix U' (2.189), the subgroups $\mathcal{G}^{\rm star}_{\rm 16vert}$ and $\mathcal{G}^{\rm circ}_{\rm 16vert}$ (as well as the semigroups $\mathcal{S}^{\rm star}_{\rm 16vert}$ and $\mathcal{S}^{\rm circ}_{\rm 16vert}$) are isomorphic by the obvious isomorphism

$$M'_{\rm star} \mapsto U' M'_{\rm circ} U'^{-1}, \quad (2.193)$$

where U' is in (2.189).

The interaction between $\mathsf{M}'_{\rm star}$ and $\mathsf{M}'_{\rm circ}$ also differs from the binary case (2.82), because

$$\mathsf{M}'_{\rm star}\mathsf{M}'_{\rm circ} = \mathsf{M}'_{\rm quad}, \quad \mathsf{M}'_{\rm circ}\mathsf{M}'_{\rm star} = \mathsf{M}'_{\rm quad}, \quad (2.194)$$

$$\mathsf{M}'_{\rm quad}\mathsf{M}'_{\rm quad} = \mathsf{M}'_{\rm quad}, \quad (2.195)$$

where $\mathsf{M}'_{\rm quad}$ is a set of 32-vertex so-called *quad-matrices* of the form

$$M'_{\text{quad}} = \begin{pmatrix} x_1 & 0 & y_1 & 0 & 0 & z_1 & 0 & s_1 \\ 0 & t_1 & 0 & u_1 & v_1 & 0 & w_1 & 0 \\ a_1 & 0 & b_1 & 0 & 0 & c_1 & 0 & d_1 \\ 0 & f_1 & 0 & g_1 & h_1 & 0 & p_1 & 0 \\ 0 & x_2 & 0 & y_2 & z_2 & 0 & s_2 & 0 \\ t_2 & 0 & u_2 & 0 & 0 & v_2 & 0 & w_2 \\ 0 & a_2 & 0 & b_2 & c_2 & 0 & d_2 & 0 \\ f_2 & 0 & g_2 & 0 & 0 & h_2 & 0 & p_2 \end{pmatrix}. \tag{2.196}$$

Because of (2.195), the set $\mathsf{M}'_{\text{quad}}$ is closed with respect to matrix multiplication, and therefore (for invertible matrices M'_{quad}) the group $\mathcal{G}^{\text{quad}}_{32\text{vert}} = \left\langle \mathsf{M}'_{\text{quad}} \middle| \cdot, (\_\_)^{-1}, \mathcal{I}_8 \right\rangle$ is a subgroup of $\mathrm{GL}(8, \mathbb{C})$. So, in trying to find higher 32-vertex solutions (having at most half as many unknown variables as a general 8×8 matrix) to the ternary braid equations (2.134), it is worthwhile to search within the class of quad-matrices (2.196).

Innovation 2.12. *The group structure of the above 16-vertex 8×8 matrices (2.192)–(2.195) is considerably different to that of 8-vertex 4×4 matrices (2.77)–(2.78) because the former contains two isomorphic binary subgroups $\mathcal{G}^{\text{star}}_{16\text{vert}}$ and $\mathcal{G}^{\text{circ}}_{16\text{vert}}$ of* $\mathrm{GL}(8, \mathbb{C})$ *(cf (2.81)–(2.83) and (2.192)).*

The sets $\mathsf{M}'_{\text{star}}$, $\mathsf{M}'_{\text{circ}}$ and $\mathsf{M}'_{\text{quad}}$ are also closed with respect to matrix addition, and therefore (because of the distributivity of $\mathbb{C}$) they are the matrix rings $\mathcal{R}^{\text{star}}_{16\text{vert}}$, $\mathcal{R}^{\text{circ}}_{16\text{vert}}$ and $\mathcal{R}^{\text{quad}}_{32vert}$, respectively. In the invertible case (2.188) and $\det M'_{\text{quad}} \neq 0$, these become matrix fields.

2.3.5 Pauli matrix presentation of the star and circle 16-vertex constant matrices

The main peculiarity of the 16-vertex 8×8 matrices (2.192)–(2.195) is the fact that they can be expressed as special tensor (Kronecker) products of the Pauli matrices (see, also, Khaneja and Glaser 2001, Ballard and Wu 2011b). Indeed, let

$$\Sigma_{ijk} = \rho_i \otimes_{\mathrm{K}} \rho_j \otimes_{\mathrm{K}} \rho_k, \qquad i, j, k = 1, 2, 3, 4, \tag{2.197}$$

where ρ_i are Pauli matrices (we have already used the letter 'σ' for the braid group generators (2.5))

$$\rho_1 = \begin{pmatrix} 0 & 1 \\ 1 & 0 \end{pmatrix}, \ \rho_2 = \begin{pmatrix} 0 & -i \\ i & 0 \end{pmatrix}, \ \rho_3 = \begin{pmatrix} 1 & 0 \\ 0 & -1 \end{pmatrix}, \ \rho_4 = I_2 = \begin{pmatrix} 1 & 0 \\ 0 & 1 \end{pmatrix}. \tag{2.198}$$

Among the total of 64 8×8 matrices Σ_{ijk} (2.197), there are 24 which generate M'_{star} (2.184) and M'_{circ} (2.185):

- Eight diagonal matrices: $\Sigma_{\text{diag}} = \{\Sigma_{333}, \Sigma_{334}, \Sigma_{343}, \Sigma_{344}, \Sigma_{433}, \Sigma_{434}, \Sigma_{443}, \Sigma_{444}\}$;

- Eight antidiagonal matrices: $\Sigma_{\text{adiag}} = \{\Sigma_{111}, \Sigma_{112}, \Sigma_{121}, \Sigma_{122}, \Sigma_{211}, \Sigma_{212}, \Sigma_{221}, \Sigma_{222}\}$;
- Eight circle-like matrices (M'_{circ} with 0s on the diagonal): $\Sigma_{\text{circ}} = \{\Sigma_{131}, \Sigma_{132}, \Sigma_{141}, \Sigma_{142}, \Sigma_{231}, \Sigma_{232}, \Sigma_{241}, \Sigma_{242}\}$.

Thus, in general we have the following set structure for the star and circle 16-vertex matrices (2.184) and (2.185)

$$\mathsf{M}'_{\text{star}} = \Sigma_{\text{diag}} \bigcup \Sigma_{\text{adiag}}, \tag{2.199}$$

$$\mathsf{M}'_{\text{circ}} = \Sigma_{\text{diag}} \bigcup \Sigma_{\text{circ}}, \tag{2.200}$$

$$\mathsf{M}'_{\text{star}} \bigcap \mathsf{M}'_{\text{circ}} = \Sigma_{\text{diag}}. \tag{2.201}$$

In particular, for the 8-vertex permutation solutions (2.144)–(2.145) of the ternary braid equations (2.134), we have

$$\tilde{c}^{\text{bisymm1, 2}}_{\text{rank}=8} = \frac{1}{2}(\Sigma_{111} + \Sigma_{444} \pm \Sigma_{212} \pm \Sigma_{343}), \tag{2.202}$$

$$\tilde{c}^{\text{symm1, 2}}_{\text{rank}=8} = \frac{1}{2}(\Sigma_{141} + \Sigma_{444} \pm \Sigma_{232} \pm \Sigma_{333}). \tag{2.203}$$

The noninvertible 16-vertex solutions M'_{star} (2.184) and M'_{circ} (2.185) having 1s on nonzero places are of *rank* = 4 and can be presented by (2.197) as follows

$$M'_{\text{star}}(1) = \begin{pmatrix} 1 & 0 & 0 & 0 & 0 & 0 & 0 & 1 \\ 0 & 1 & 0 & 0 & 0 & 0 & 1 & 0 \\ 0 & 0 & 1 & 0 & 0 & 1 & 0 & 0 \\ 0 & 0 & 0 & 1 & 1 & 0 & 0 & 0 \\ 0 & 0 & 0 & 1 & 1 & 0 & 0 & 0 \\ 0 & 0 & 1 & 0 & 0 & 1 & 0 & 0 \\ 0 & 1 & 0 & 0 & 0 & 0 & 1 & 0 \\ 1 & 0 & 0 & 0 & 0 & 0 & 0 & 1 \end{pmatrix} = \Sigma_{111} + \Sigma_{444}, \tag{2.204}$$

$$M'_{\text{circ}}(1) = \begin{pmatrix} 1 & 0 & 0 & 0 & 0 & 1 & 0 & 0 \\ 0 & 1 & 0 & 0 & 1 & 0 & 0 & 0 \\ 0 & 0 & 1 & 0 & 0 & 0 & 0 & 1 \\ 0 & 0 & 0 & 1 & 0 & 0 & 1 & 0 \\ 0 & 1 & 0 & 0 & 1 & 0 & 0 & 0 \\ 1 & 0 & 0 & 0 & 0 & 1 & 0 & 0 \\ 0 & 0 & 0 & 1 & 0 & 0 & 1 & 0 \\ 0 & 0 & 1 & 0 & 0 & 0 & 0 & 1 \end{pmatrix} = \Sigma_{141} + \Sigma_{444}. \tag{2.205}$$

Similarly, one can obtain the Pauli matrix presentation for the general star and circle 16-vertex matrices (2.184) and (2.185) which will contain linear combinations of the 16 parameters as coefficients before the Σs.

2.3.6 Invertible and noninvertible 16-vertex solutions to the ternary braid equations

First, consider the 16-vertex solutions to (2.134) having the star matrix shape (2.184). We found the following two one-parameter invertible solutions

$$\tilde{c}^{16-\text{vert,star}}_{\text{rank}=8}(x) = \begin{pmatrix} x^3 & 0 & 0 & 0 & 0 & 0 & 0 & -1 \\ 0 & x^3 & 0 & 0 & 0 & 0 & \mp x^2 & 0 \\ 0 & 0 & x^3 & 0 & 0 & -x^2 & 0 & 0 \\ 0 & 0 & 0 & x^3 & \mp x^4 & 0 & 0 & 0 \\ 0 & 0 & 0 & \pm x^2 & x^3 & 0 & 0 & 0 \\ 0 & 0 & x^4 & 0 & 0 & x^3 & 0 & 0 \\ 0 & \pm x^4 & 0 & 0 & 0 & 0 & x^3 & 0 \\ x^6 & 0 & 0 & 0 & 0 & 0 & 0 & x^3 \end{pmatrix}, \tag{2.206}$$

$$\text{tr}\,\tilde{c} = 8x^3,$$

$$\det \tilde{c} = 16x^{24}, \quad x \neq 0,$$

$$\text{eigenvalues:} \left\{(1+i)x^3\right\}^{[4]}, \left\{(1-i)x^3\right\}^{[4]}.$$

Both matrices in (2.206) are diagonalizable and are conjugates via

$$U_{\text{star}} = \begin{pmatrix} 1 & 0 & 0 & 0 & 0 & 0 & 0 & 0 \\ 0 & -1 & 0 & 0 & 0 & 0 & 0 & 0 \\ 0 & 0 & 1 & 0 & 0 & 0 & 0 & 0 \\ 0 & 0 & 0 & -1 & 0 & 0 & 0 & 0 \\ 0 & 0 & 0 & 0 & 1 & 0 & 0 & 0 \\ 0 & 0 & 0 & 0 & 0 & 1 & 0 & 0 \\ 0 & 0 & 0 & 0 & 0 & 0 & 1 & 0 \\ 0 & 0 & 0 & 0 & 0 & 0 & 0 & 1 \end{pmatrix}, \tag{2.207}$$

which cannot be presented in the form of a triple Kronecker product (2.141). Therefore, the two solutions in (2.206) are not q-conjugate and become different 16-vertex one-parameter invertible solutions of the braid equations (2.134).

In search of 16-vertex solutions to the total braid equations (2.134) of the circle matrix shape (2.185), we found that only noninvertible ones exist. They are the following two 2-parameter solutions of rank 4

$$\tilde{c}^{16-\text{vert,circ}}_{\text{rank}=4}(x, y) = \begin{pmatrix} \pm xy & 0 & 0 & 0 & 0 & y^2 & 0 & 0 \\ 0 & \pm xy & 0 & 0 & xy & 0 & 0 & 0 \\ 0 & 0 & \pm xy & 0 & 0 & 0 & 0 & y^2 \\ 0 & 0 & 0 & \pm xy & 0 & 0 & xy & 0 \\ 0 & xy & 0 & 0 & \pm xy & 0 & 0 & 0 \\ x^2 & 0 & 0 & 0 & 0 & \pm xy & 0 & 0 \\ 0 & 0 & 0 & xy & 0 & 0 & \pm xy & 0 \\ 0 & 0 & x^2 & 0 & 0 & 0 & 0 & \pm xy \end{pmatrix}, \tag{2.208}$$

$$\text{tr}\,\tilde{c} = \pm 8xy,$$

$$\text{eigenvalues: } \{2xy\}^{[4]}, \{0\}^{[4]}.$$

Two matrices in (2.208) are not even conjugates in the standard way, and so they are different 16-vertex two-parameter noninvertible solutions to the braid equations (2.134).

For the only partial 13-braid equation (2.136), there are four polynomial 16-vertex two-parameter invertible solutions

$$\tilde{c}_{\text{rank}=8}^{16-\text{vert},\ 13\text{circ}}(x,y) = \begin{pmatrix} x & 0 & 0 & 0 & 0 & y^2 & 0 & 0 \\ 0 & xy & 0 & 0 & x & 0 & 0 & 0 \\ 0 & 0 & x & 0 & 0 & 0 & 0 & \pm y^2 \\ 0 & 0 & 0 & xy & 0 & 0 & \pm x & 0 \\ 0 & x & 0 & 0 & xy & 0 & 0 & 0 \\ x^2 & 0 & 0 & 0 & 0 & x & 0 & 0 \\ 0 & 0 & 0 & \pm x & 0 & 0 & xy & 0 \\ 0 & 0 & \pm x^2 & 0 & 0 & 0 & 0 & x \end{pmatrix}, \tag{2.209}$$

$$\begin{pmatrix} x & 0 & 0 & 0 & 0 & y^2 & 0 & 0 \\ 0 & xy & 0 & 0 & -x & 0 & 0 & 0 \\ 0 & 0 & x & 0 & 0 & 0 & 0 & \pm y^2 \\ 0 & 0 & 0 & xy & 0 & 0 & \mp x & 0 \\ 0 & -x & 0 & 0 & xy & 0 & 0 & 0 \\ x^2 & 0 & 0 & 0 & 0 & x & 0 & 0 \\ 0 & 0 & 0 & \mp x & 0 & 0 & xy & 0 \\ 0 & 0 & \pm x^2 & 0 & 0 & 0 & 0 & x \end{pmatrix}, \tag{2.210}$$

$$\begin{gathered} \operatorname{tr} \tilde{c} = 4x(y+1), \\ \det \tilde{c} = x^8(y^2-1)^4,\ x \neq 0,\ y \neq 1, \\ \text{eigenvalues}: \{x(y+1)\}^{[4]},\ \{x(y-1)\}^{[2]},\ \{-x(y-1)\}^{[2]}. \end{gathered} \tag{2.211}$$

Also, for the partial 13-braid equation (2.136), we found four exotic irrational (an analog of (2.89) for the Yang–Baxter equation (2.12)) 16-vertex, two-parameter invertible solutions of rank 8 of the form

$$\tilde{c}_{\text{rank}=8}^{16-\text{vert},\ 13\text{circ},\ 1}(x,y) = \begin{pmatrix} x(2y-1) & 0 & 0 & 0 & 0 & y^2 & 0 & 0 \\ 0 & xy & 0 & 0 & x\sqrt{2(y-1)y+1} & 0 & 0 & 0 \\ 0 & 0 & x(2y-1) & 0 & 0 & 0 & 0 & \pm y^2 \\ 0 & 0 & 0 & xy & 0 & 0 & \pm x\sqrt{2(y-1)y+1} & 0 \\ 0 & x\sqrt{2(y-1)y+1} & 0 & 0 & xy & 0 & 0 & 0 \\ x^2 & 0 & 0 & 0 & 0 & x & 0 & 0 \\ 0 & 0 & 0 & \pm x\sqrt{2(y-1)y+1} & 0 & 0 & xy & 0 \\ 0 & 0 & \pm x^2 & 0 & 0 & 0 & 0 & x \end{pmatrix}, \tag{2.212}$$

$$\tilde{c}^{16-\text{vert},\,13\text{circ},\,2}_{\text{rank}=8}(x,y)$$

$$= \begin{pmatrix} x(2y-1) & 0 & 0 & 0 & 0 & y^2 & 0 & 0 \\ 0 & xy & 0 & 0 & -x\sqrt{2(y-1)y+1} & 0 & 0 & 0 \\ 0 & 0 & x(2y-1) & 0 & 0 & 0 & 0 & \pm y^2 \\ 0 & 0 & 0 & xy & 0 & 0 & \mp x\sqrt{2(y-1)y+1} & 0 \\ 0 & -x\sqrt{2(y-1)y+1} & 0 & 0 & xy & 0 & 0 & 0 \\ x^2 & 0 & 0 & 0 & 0 & x & 0 & 0 \\ 0 & 0 & 0 & \mp x\sqrt{2(y-1)y+1} & 0 & 0 & xy & 0 \\ 0 & 0 & \pm x^2 & 0 & 0 & 0 & 0 & x \end{pmatrix}, \tag{2.213}$$

$$\operatorname{tr}\tilde{c} = 8xy,\quad \det\tilde{c} = x^8(y-1)^8,\qquad x \neq 0,\quad y \neq 1, \tag{2.214}$$

$$\text{eigenvalues}: \left\{x\left(y+\sqrt{2(y-1)y+1}\right)\right\}^{[4]}, \left\{x\left(y-\sqrt{2(y-1)y+1}\right)\right\}^{[4]}. \tag{2.215}$$

The matrices in (2.209)–(2.213) are diagonalizable, have the same eigenvalues (2.215), and are pairwise conjugate by

$$U_{\text{circ}} = \begin{pmatrix} 1 & 0 & 0 & 0 & 0 & 0 & 0 & 0 \\ 0 & 1 & 0 & 0 & 0 & 0 & 0 & 0 \\ 0 & 0 & -1 & 0 & 0 & 0 & 0 & 0 \\ 0 & 0 & 0 & -1 & 0 & 0 & 0 & 0 \\ 0 & 0 & 0 & 0 & 1 & 0 & 0 & 0 \\ 0 & 0 & 0 & 0 & 0 & 1 & 0 & 0 \\ 0 & 0 & 0 & 0 & 0 & 0 & 1 & 0 \\ 0 & 0 & 0 & 0 & 0 & 0 & 0 & 1 \end{pmatrix}. \tag{2.216}$$

Because U_{circ} cannot be presented in the form (2.141), all solutions in (2.209)–(2.213) are not mutually q-conjugate and become eight different 16-vertex two-parameter invertible solutions to the partial 13-braid equation (2.136). If $y = 1$, then the matrices (2.209)–(2.213) become of rank 4 with vanishing determinants (2.211), (2.214), and therefore in this case they are a 16-vertex one-parameter circle of noninvertible solutions to the total braid equations (2.134).

Further families of solutions could be constructed using additional parameters: the scaling parameter t in (2.139) and the complex elements of the matrix q (2.140).

2.3.7 Higher 2^n-vertex constant solutions to n-ary braid equations

Next we considered the 4-ary constant braid equations (2.114)–(2.116) and found the following 32-vertex star solution

$$\tilde{c}_{16} = \begin{pmatrix} 1&0&0&0&0&0&0&0&0&0&0&0&0&0&0&-1\\ 0&1&0&0&0&0&0&0&0&0&0&0&0&0&-1&0\\ 0&0&1&0&0&0&0&0&0&0&0&0&0&-1&0&0\\ 0&0&0&1&0&0&0&0&0&0&0&0&-1&0&0&0\\ 0&0&0&0&1&0&0&0&0&0&0&-1&0&0&0&0\\ 0&0&0&0&0&1&0&0&0&0&-1&0&0&0&0&0\\ 0&0&0&0&0&0&1&0&0&-1&0&0&0&0&0&0\\ 0&0&0&0&0&0&0&1&-1&0&0&0&0&0&0&0\\ 0&0&0&0&0&0&0&1&1&0&0&0&0&0&0&0\\ 0&0&0&0&0&0&1&0&0&1&0&0&0&0&0&0\\ 0&0&0&0&0&1&0&0&0&0&1&0&0&0&0&0\\ 0&0&0&0&1&0&0&0&0&0&0&1&0&0&0&0\\ 0&0&0&1&0&0&0&0&0&0&0&0&1&0&0&0\\ 0&0&1&0&0&0&0&0&0&0&0&0&0&1&0&0\\ 0&1&0&0&0&0&0&0&0&0&0&0&0&0&1&0\\ 1&0&0&0&0&0&0&0&0&0&0&0&0&0&0&1 \end{pmatrix}. \quad (2.217)$$

We may compare (2.217) with particular cases of the star solutions to the Yang–Baxter equation (2.87) and the ternary braid equation (2.206)

$$\tilde{c}_4 = \begin{pmatrix} 1&0&0&-1\\ 0&1&-1&0\\ 0&1&1&0\\ 1&0&0&1 \end{pmatrix}, \quad \tilde{c}_8 = \begin{pmatrix} 1&0&0&0&0&0&0&-1\\ 0&1&0&0&0&0&-1&0\\ 0&0&1&0&0&-1&0&0\\ 0&0&0&1&-1&0&0&0\\ 0&0&0&1&1&0&0&0\\ 0&0&1&0&0&1&0&0\\ 0&1&0&0&0&0&1&0\\ 1&0&0&0&0&0&0&1 \end{pmatrix}. \quad (2.218)$$

Informally we call such solutions the Minkowski star solutions because their legs have the Minkowski signature. Thus, we assume that in the general case for the n-ary braid equation there exist 2^{n+1}-vertex $2^n \times 2^n$ matrix Minkowski star invertible solutions of the above form

$$\tilde{c}_{2^n} = \begin{pmatrix} 1&0&0&0&0&-1\\ 0&\ddots&0&0&⋰&0\\ 0&0&1&-1&0&0\\ 0&0&1&1&0&0\\ 0&⋰&0&0&\ddots&0\\ 1&0&0&0&0&1 \end{pmatrix}. \quad (2.219)$$

This allows us to use the general solution (2.219) as n-ary braiding quantum gates with an arbitrary number of qubits.

2.4 Invertible and noninvertible quantum gates

Informally, quantum computing consists of preparation (setting up an initial quantum state), evolution (by a quantum circuit), and measurement (projection onto the final state). Mathematically (in the computational basis), the initial state is a vector in a Hilbert space (multi-qubit state), the evolution is governed by successive (quantum circuit) invertible linear transformations (unitary matrices called quantum gates), and the measurement is made by noninvertible projection matrices to leave only one final quantum (multi-qubit) state. So, quantum computing is noninvertible overall, and we may consider noninvertible transformations at each step. It was then realized that one can invite the Yang–Baxter operators (solutions of the constant Yang–Baxter equation) into quantum gates, providing a means of entangling otherwise non-entangled states. This insight uncovered a deep connection between quantum and topological computation (for details, see e.g. Kauffman and Lomonaco 2002, 2004).

Here we propose extending the above picture in two directions. First, we can treat higher braided operators as higher braiding gates. Second, we will analyze the possible role of noninvertible linear transformations (described by the partial unitary matrices introduced in (2.20)–(2.21)), which can be interpreted as a property of some hypothetical quantum circuit, e.g., with specific loss of information, some kind of dissipativity or vagueness. This can be considered as an intermediate case between standard unitary computing and the measurement only computing of Bonderson *et al* (2008).

To establish notation recall (Nielsen and Chuang 2000) that in the computational basis (vector representation) and Dirac notation, a (pure) one-qubit state is described by a vector in two-dimensional Hilbert space $V = \mathbb{C}^2$

$$\begin{gathered} |\psi\rangle \equiv |\psi^{(1)}\rangle = a_0|0\rangle + a_1|1\rangle, \quad |0\rangle = \begin{pmatrix}1\\0\end{pmatrix}, \quad |1\rangle = \begin{pmatrix}0\\1\end{pmatrix}, \\ |a_0|^2 + |a_1|^2 = 1, \qquad a_i \in \mathbb{C}, \quad i = 1, 2, \end{gathered} \tag{2.220}$$

where a_i is a probability amplitude of $|i\rangle$. Sometimes, for a one-qubit state it is convenient to use the Bloch unit sphere representation (normalized up to a general unimportant and unmeasurable phase)

$$|\psi(\theta, \gamma)\rangle = \cos\frac{\theta}{2}|0\rangle + e^{i\gamma}\sin\frac{\theta}{2}|1\rangle, \quad 0 \leqslant \theta \leqslant \pi, \quad 0 \leqslant \gamma \leqslant 2\pi. \tag{2.221}$$

A (pure) state of L-qubits $|\psi^{(L)}\rangle$ is described by 2^L amplitudes, and so is a vector in 2^L-dimensional Hilbert space. If $|\psi^{(L)}\rangle$ cannot be presented as a tensor product of L one-qubit states (2.220), then it is called *entangled*. For instance, a two-qubit pure state

$$\begin{aligned} &|\psi^{(2)}\rangle = a_{00}|00\rangle + a_{01}|01\rangle + a_{10}|10\rangle + a_{11}|11\rangle, \\ &|a_{00}|^2 + |a_{01}|^2 + |a_{10}|^2 + |a_{11}|^2 = 1, \\ &a_{ij} \in \mathbb{C}, \, i, j = 1, 2, \end{aligned} \tag{2.222}$$

is entangled, if $\det(a_{ij}) \neq 0$, and the *concurrence*

$$C^{(2)} \equiv C^{(2)}(|\,\psi^{(2)}\rangle) = 2\,\big|\det(a_{ij})\big| \tag{2.223}$$

is the measure of entanglement $0 \leqslant C^{(2)} \leqslant 1$. It follows from (2.220) that the tensor product of states has vanishing concurrence $C^{(2)}(|\,\psi_1\rangle \otimes |\,\psi_2\rangle) = 0$. An example of the maximally entangled ($C^{(2)} = 1$) two-qubit states is the (first) Bell state $|\,\psi^{(2)}\rangle_{\text{Bell}} = (|\,00\rangle + |\,11\rangle)/\sqrt{2}$.

The concurrence of the three-qubit state

$$|\,\psi^{(3)}\rangle = \sum_{i,j,k=0}^{1} a_{ijk}\,|\,ijk\rangle, \qquad \sum_{i,j,k=0}^{1} \big|\,a_{ijk}\big|^2 = 1, \quad a_{ijk} \in \mathbb{C}, \tag{2.224}$$

is determined by the Cayley's $2 \times 2 \times 2$ hyperdeterminant

$$C^{(3)} = 4\,\big|\,\text{hdet}(a_{ijk})\,\big|, \qquad 0 \leqslant C^{(3)} \leqslant 1, \tag{2.225}$$

$$\begin{aligned} \text{hdet}(a_{ijk}) &= a_{000}^2 a_{111}^2 + a_{001}^2 a_{110}^2 + a_{010}^2 a_{101}^2 + a_{100}^2 a_{011}^2 - 2a_{000}a_{001}a_{110}a_{111} \\ &- 2a_{000}a_{010}a_{101}a_{111} - 2a_{000}a_{011}a_{100}a_{111} - 2a_{001}a_{010}a_{101}a_{111} - 2a_{001}a_{011}a_{100}a_{110} \\ &- 2a_{010}a_{011}a_{100}a_{101} + 4a_{000}a_{011}a_{101}a_{110} + 4a_{001}a_{010}a_{100}a_{111}. \end{aligned} \tag{2.226}$$

Thus, if the three-qubit state (2.224) is not entangled, then $C^{(3)} = 0$ (for the tensor product of one-qubit states). One of the maximally entangled ($C^{(3)} = 1$) three-qubit states is the GHZ state $|\,\psi^{(3)}\rangle_{\text{GHZ}} = (|\,000\rangle + |\,111\rangle)/\sqrt{2}$.

A quantum L-qubit gate is a linear transformation of 2^L-dimensional Hilbert space $(\mathbb{C}^2)^{\otimes L} \to (\mathbb{C}^2)^{\otimes L}$ which in the computational basis (2.220) is described of the $2^L \times 2^L$ matrix $U^{(L)}$ such that the L-qubit state transforms as $|\,\psi^{'(L)}\rangle = U^{(L)}\,|\,\psi^{(L)}\rangle$. In this way, a *quantum circuit* is described as the successive application of elementary gates to an initial quantum state, which is the product of the corresponding matrices (for details, see, e.g., Nielsen and Chuang 2000). It is a standard assumption that each elementary L-qubit transformation is *unitary*, which implies the following strong restriction on the corresponding matrix $U \equiv U^{(L)}$ as

$$U^\star U = UU^\star = I \equiv I_{2^L \times 2^L}, \tag{2.227}$$

where I is the $2^L \times 2^L$ identity matrix for L-qubit state and the operation $(\star)$ is the conjugate-transposition. The first equality in (2.227) means that the matrix $U^{(L)}$ is *normal* (cf (2.20)–(2.21)). The equations (2.227) follow from the definition of the *adjoint operator*

$$\langle U\psi^{(L)} | I\varphi^{(L)}\rangle = \langle I\psi^{(L)} | U^\star \varphi^{(L)}\rangle \tag{2.228}$$

applied to this simplest example of L-qubits in the 2^L-dimensional Hilbert space $(\mathbb{C}^2)^{\otimes L}$ (for the general case the derivation almost literally coincides), which we write in the following special form (in Dirac notation with bra- and ket- vectors) with explicitly added identities. Then, (2.227) follows from (2.228) as

$$\langle U^\star U\psi^{(L)}|I\varphi^{(L)}\rangle = \langle I\psi^{(L)}|UU^\star\varphi^{(L)}\rangle = \langle I\psi^{(L)}|I\varphi^{(L)}\rangle, \tag{2.229}$$

and any unitary matrix preserves the inner product

$$\langle U\psi^{(L)}|U\varphi^{(L)}\rangle = \langle I\psi^{(L)}|I\varphi^{(L)}\rangle, \tag{2.230}$$

which means that unitary operators satisfying (2.227) are bounded operators (bounded matrices in our case) and invertible with the inverse $U^{-1} = U^\star$.

Let us consider a possibility of noninvertible intermediate transformations of L-qubit states, i.e., *noninvertible gates*, which are described by the $2^L \times 2^L$ matrices $U(r)$ of (possibly) less than full rank $1 \leqslant r \leqslant 2^L$. This can be related to the production of degenerate states (see, e.g. Jaffali and Oeding 2020), particle loss (Neven *et al* 2018, Fraïsse and Braun 2016, Zangi and Qiao 2021), and the role of ranks in multiparticle entanglement (Chong *et al* 2005, Bruzda *et al* 2019).

In the limited cases $U(r = 2^L) \equiv U = U^{(L)}$, and $U(1)$ corresponds to the measurement matrix being the projection to one final vector $|\, i_{\text{final}}\rangle$. In this case, for noninvertible transformations with $r < 2^L$ instead of unitarity (2.227), we consider partial unitarity (2.20)–(2.21) as

$$U(r)^\star U(r) = I_1(r), \tag{2.231}$$

$$U(r)U(r)^\star = I_2(r), \tag{2.232}$$

where $I_1(r)$ and $I_2(r)$ are (or may be) different partial shuffle identities having r units on the diagonal. There is an exotic limiting case, which is impossible for the identity I: we call two partial identities *orthogonal*, if

$$I_1(r)I_2(r) = Z, \tag{2.233}$$

where $Z = Z_{2^L\times 2^L}$ is the zero $2^L \times 2^L$ matrix.

We propose corresponding noninvertible analogs of (2.228)–(2.230) as follows. The *partial adjoint operator* $U(r)^\star$ in the 2^L-dimensional Hilbert space $(\mathbb{C}^2)^{\otimes L}$ is defined by

$$\langle U(r)\psi^{(L)}|I_2(r)\varphi^{(L)}\rangle = \langle I_1(r)\psi^{(L)}|U(r)^\star\varphi^{(L)}\rangle, \tag{2.234}$$

such that (see (2.231)–(2.232))

$$\langle U(r)^\star U(r)\psi^{(L)}|I_2(r)\varphi^{(L)}\rangle = \langle I_1(r)\psi^{(L)}|U(r)U(r)^\star\varphi^{(L)}\rangle = \langle I_1(r)\psi^{(L)}|I_2(r)\varphi^{(L)}\rangle. \tag{2.235}$$

We call the rhs of (2.235) the *partial inner product*. So instead of (2.230), we define $U(r)$ as the *partially bounded operator*

$$\langle U(r)\psi^{(L)}|U(r)\varphi^{(L)}\rangle = \langle I_1(r)\psi^{(L)}|I_2(r)\varphi^{(L)}\rangle. \tag{2.236}$$

Thus, if the partial identities are orthogonal (2.233), then the partial inner product vanishes identically, and the operator $U(r)$ becomes a zero norm operator in the sense of (2.236), although (2.231)–(2.232) are not zero.

In case the rank r is fixed, there can be $(2^L!/r!(2^L - r)!)^2$ partial unitary matrices $U(r)$ satisfying (2.231)–(2.232).

We define a *general unitary semigroup* as a semigroup of matrices $U(r)$ of rank r satisfying partial regularity (2.231)–(2.232) (in the symmetric case $I_1(r) = I_2(r) \equiv I(r)$).

As an example, we consider two 2-qubit states (2.222) $|\psi^{(2)}\rangle$ and $|\varphi^{(2)}\rangle$ (with a'_{ij} and $|i'j'\rangle$) and the noninvertible transformation described by three-parameter 4×4 matrices of rank 3 (but which are not nilpotent)

$$U(3) = U^{(L=2)}(r=3) = \begin{pmatrix} 0 & 0 & 0 & 0 \\ 0 & e^{i\beta} & 0 & 0 \\ 0 & 0 & 0 & e^{i\gamma} \\ e^{i\alpha} & 0 & 0 & 0 \end{pmatrix}, \quad \alpha, \beta, \gamma \in \mathbb{R}. \tag{2.237}$$

The partial unitarity (2.231)–(2.232) and partial identities now become

$$U(3)^\star U(3) = I_1(3) = \begin{pmatrix} 1 & 0 & 0 & 0 \\ 0 & 1 & 0 & 0 \\ 0 & 0 & 0 & 0 \\ 0 & 0 & 0 & 1 \end{pmatrix}, \tag{2.238}$$

$$U(3)U(3)^\star = I_2(3) = \begin{pmatrix} 0 & 0 & 0 & 0 \\ 0 & 1 & 0 & 0 \\ 0 & 0 & 1 & 0 \\ 0 & 0 & 0 & 1 \end{pmatrix}. \tag{2.239}$$

The partial identities (2.238)–(2.239) are not orthogonal (2.233), because

$$I_1(3)I_2(3) = \begin{pmatrix} 0 & 0 & 0 & 0 \\ 0 & 1 & 0 & 0 \\ 0 & 0 & 0 & 0 \\ 0 & 0 & 0 & 1 \end{pmatrix} \neq Z, \tag{2.240}$$

which directly gives the signature of the partial inner product (2.235), in our case of the Hilbert space $(\mathbb{C}^2)^{\otimes 2}$.

The definition of a partial adjoint operator (2.234) is satisfied with both sides being equal to $a_{00}a'_{11}e^{i\alpha}\langle 00|1'1'\rangle + a_{01}a'_{01}e^{i\beta}\langle 01|0'1'\rangle + a_{11}a'_{10}e^{i\gamma}\langle 11|1'0'\rangle$. The partial boundedness condition (2.236) holds with the partial inner product (2.235) becoming $a_{01}a'_{01}\langle 01|0'1'\rangle + a_{11}a'_{11}\langle 11|1'1'\rangle$, thus $U(3)$ (2.237), which is a bounded partial unitary operator.

An example of a zero norm (in our sense (2.236)) operator is the two-parameter partial unitary rank 2 matrix

$$U_{\mathrm{nil}}(2) = U^{(L=2)}(r=2) = \begin{pmatrix} 0 & 0 & 0 & 0 \\ 0 & 0 & 0 & e^{i\beta} \\ e^{i\alpha} & 0 & 0 & 0 \\ 0 & 0 & 0 & 0 \end{pmatrix}, \quad \alpha, \beta \in \mathbb{R}. \tag{2.241}$$

The partial unitarity relations for $U_{\mathrm{nil}}(2)$ have the form

$$U_{\rm nil}(2)^{\star}\,U_{\rm nil}(2) = I_{\rm nil,1}(2) = \begin{pmatrix} 1 & 0 & 0 & 0 \\ 0 & 0 & 0 & 0 \\ 0 & 0 & 0 & 0 \\ 0 & 0 & 0 & 1 \end{pmatrix}, \tag{2.242}$$

$$U_{\rm nil}(2)\,U_{\rm nil}(2)^{\star} = I_{\rm nil,2}(2) = \begin{pmatrix} 0 & 0 & 0 & 0 \\ 0 & 1 & 0 & 0 \\ 0 & 0 & 1 & 0 \\ 0 & 0 & 0 & 0 \end{pmatrix}. \tag{2.243}$$

It may be seen that the partial identities $I_{\rm nil,1}(2)$ and $I_{\rm nil,2}(2)$ are now orthogonal (2.233), and the partial inner product (2.235) vanishes identically, and also the boundedness condition (2.236) holds with the rhs vanishing, despite $U_{\rm nil}(2)$ being a nonzero nilpotent matrix (2.241).

2.5 Binary braiding quantum gates

Let us consider those Yang–Baxter maps that could be linear transformations of two-qubit spaces. We will pay attention to the most general 8-vertex solutions to the Yang–Baxter equations (2.87)–(2.94) and (2.97)–(2.98), which are unitary (and invertible) or partial unitary (2.20)–(2.21) (and noninvertible).

Consider the unitary version of the invertible star 8-vertex solutions (2.87)–(2.89) to the matrix Yang–Baxter equation (2.12). We use the exponential form of the parameters

$$\begin{gathered} x = r_x e^{i\alpha},\ \ y = r_y e^{i\beta},\ \ z = r_z e^{i\gamma}, \\ r_{x,y,z},\ \alpha,\ \beta,\ \gamma \in \mathbb{R},\quad r_{x,y,z} \geqslant 0,\quad |\,\alpha\,|,\,|\,\beta\,|,\,|\,\gamma\,| \leqslant 2\pi. \end{gathered} \tag{2.244}$$

For (2.87), exploiting unitarity (2.227) we obtain

$$U^{8-\text{vert,star}}_{\text{rank}=4}(\alpha,\beta) = \frac{1}{\sqrt{2}} \begin{pmatrix} e^{i(\alpha+\beta)} & 0 & 0 & e^{2i\beta} \\ 0 & e^{i(\alpha+\beta)} & \pm\, e^{i(\alpha+\beta)} & 0 \\ 0 & \mp\, e^{i(\alpha+\beta)} & e^{i(\alpha+\beta)} & 0 \\ -\,e^{2i\alpha} & 0 & 0 & e^{i(\alpha+\beta)} \end{pmatrix}, \tag{2.245}$$

$$\begin{aligned} \operatorname{tr} U &= 2\sqrt{2}\,e^{i(\alpha+\beta)}, \\ \det U &= e^{4i(\alpha+\beta)}, \end{aligned} \tag{2.246}$$

$$\text{eigenvalues}:\ \left\{-(-1)^{3/4}e^{i(\alpha+\beta)}\right\}^{[2]},\ \left\{(-1)^{1/4}e^{i(\alpha+\beta)}\right\}^{[2]}. \tag{2.247}$$

With the particular choice of parameters $\alpha = \beta = 0$ and lower signs, the solution (2.245) coincides with the 8-vertex braiding gate of Kauffman and Lomonaco (2004).

Next we search for unitary solutions among the invertible circle of 8-vertex traceless solutions (2.97) to the matrix Yang–Baxter equation (2.12) with parameters in the exponential form (2.244). The unitarity conditions (2.227) give the following equations on the parameters (2.244)

$$r = r_y = r_z, \quad r^2(r_x^2 + r^2) = 1, \quad r^8 + r^6 - 2r^4 + 1 = r^2 \tag{2.248}$$

$$\alpha - \beta = \frac{\pi}{2}. \tag{2.249}$$

The system of equations (2.248) has two real positive (or zero) solutions

$$(1) \quad r_x = 1, \quad r = \sqrt{\frac{\sqrt{5}-1}{2}}, \tag{2.250}$$

$$(2) \quad r_x = 0, \quad r = 1. \tag{2.251}$$

Thus, only the first solution leads to an 8-vertex two-parameter unitary braiding quantum gate of the form (we put $\gamma \mapsto \beta$ in (2.244))

$$U_{\text{rank}=4}^{8-\text{vert,circ}}(\alpha, \beta) = \sqrt{\frac{\sqrt{5}-1}{2}} \begin{pmatrix} 0 & e^{i(\alpha+\beta)} & ie^{i(\alpha+\beta)}\sqrt{\frac{\sqrt{5}-1}{2}} & 0 \\ -e^{2i\alpha}\sqrt{\frac{\sqrt{5}-1}{2}} & 0 & 0 & e^{i(\alpha+\beta)} \\ ie^{2i\alpha} & 0 & 0 & ie^{i(\alpha+\beta)}\sqrt{\frac{\sqrt{5}-1}{2}} \\ 0 & -e^{2i\alpha}\sqrt{\frac{\sqrt{5}-1}{2}} & ie^{2i\alpha} & 0 \end{pmatrix}, \tag{2.252}$$

$$\det U = e^{2i(3\alpha+\beta)}. \tag{2.253}$$

The second solution (2.251) gives 4-vertex two-parameter unitary braiding quantum gate (we also put $\gamma \mapsto \beta$ in (2.244))

$$U_{\text{rank}=4}^{4-\text{vert,circ}}(\alpha, \beta) = \begin{pmatrix} 0 & 0 & e^{i(\alpha+\beta)} & 0 \\ e^{2i\alpha} & 0 & 0 & 0 \\ 0 & 0 & 0 & e^{i(\alpha+\beta)} \\ 0 & e^{2i\alpha} & 0 & 0 \end{pmatrix}, \quad \det U = -e^{2i(3\alpha+\beta)}. \tag{2.254}$$

The noninvertible 8-vertex circle solution (2.98) to the Yang–Baxter equation (2.12) cannot be partial unitary (2.231)–(2.232) with any values of its parameters.

2.6 Higher braiding quantum gates

In general, only special linear transformations of 2^L-dimensional Hilbert space can be treated as elementary quantum gates for an L-qubit state (Nielsen and Chuang 2000). First, in the invertible case, the transformations should be unitary (2.227), and in the hypothetical noninvertible case they can satisfy partial unitarity (2.231)–

(2.232). Second, the braiding gates have to be $2^L \times 2^L$ matrix solutions to the constant Yang–Baxter equation (Kauffman and Lomonaco 2004) or higher braid equations (2.114)–(2.116). Here we consider (as a lowest case higher example) the ternary braiding gates acting on 3-qubit quantum states, i.e., 8×8 matrix solutions to the ternary braid equations (2.134), which satisfy unitarity (2.227) or partial unitarity (2.231)–(2.232).

Note that all the permutation solutions (2.144)–(2.145) are by definition unitary, and are therefore ternary braiding gates automatically, and we call them *permutation8-vertex ternary braiding quantum gates* $U_{\text{perm}}^{8-\text{vertex}}$. By the same reasoning the unitary version of the invertible star 8-vertex parameter-permutation solutions (2.146)–(2.153) to the ternary braid equations (2.134) will contain the complex numbers of unit magnitude as parameters.

Indeed, for the bisymmetric series (2.146)–(2.147) of star-like solutions we have four two-real parameter unitary ternary braiding quantum gates ($\varkappa = \pm 1$)

$$U_{\text{bisymm1}}^{8-\text{vertex}}(\alpha, \beta) = \begin{pmatrix} e^{i(\alpha+\beta)} & 0 & 0 & 0 & 0 & 0 & 0 & 0 \\ 0 & 0 & 0 & 0 & 0 & 0 & \varkappa e^{2i\beta} & 0 \\ 0 & 0 & e^{i(\alpha+\beta)} & 0 & 0 & 0 & 0 & 0 \\ 0 & 0 & 0 & 0 & \varkappa e^{2i\alpha} & 0 & 0 & 0 \\ 0 & 0 & 0 & \pm\varkappa e^{2i\beta} & 0 & 0 & 0 & 0 \\ 0 & 0 & 0 & 0 & 0 & e^{i(\alpha+\beta)} & 0 & 0 \\ 0 & \pm\varkappa e^{2i\alpha} & 0 & 0 & 0 & 0 & 0 & 0 \\ 0 & 0 & 0 & 0 & 0 & 0 & 0 & e^{i(\alpha+\beta)} \end{pmatrix}, \tag{2.255}$$
$$\alpha, \beta \in \mathbb{R}, \quad |\alpha|, |\beta| \leqslant 2\pi,$$

which is a ternary analog of the first parameter-permutation solution to the Yang–Baxter equation from (2.33). The ternary analog of the second star solution is the following unitary version of the bisymmetric series (2.148)–(2.149)

$$U_{\text{bisymm2}}^{8-\text{vertex}}(\alpha, \beta) = \begin{pmatrix} 0 & 0 & 0 & 0 & 0 & 0 & 0 & e^{6i\alpha} \\ 0 & \varkappa e^{3i(\alpha+\beta)} & 0 & 0 & 0 & 0 & 0 & 0 \\ 0 & 0 & 0 & 0 & 0 & e^{2i(2\alpha+\beta)} & 0 & 0 \\ 0 & 0 & 0 & \varkappa e^{3i(\alpha+\beta)} & 0 & 0 & 0 & 0 \\ 0 & 0 & 0 & 0 & \varkappa e^{3i(\alpha+\beta)} & 0 & 0 & 0 \\ 0 & 0 & \pm e^{2i(\alpha+2\beta)} & 0 & 0 & 0 & 0 & 0 \\ 0 & 0 & 0 & 0 & 0 & 0 & \varkappa e^{3i(\alpha+\beta)} & 0 \\ \pm e^{6i\beta} & 0 & 0 & 0 & 0 & 0 & 0 & 0 \end{pmatrix}. \tag{2.256}$$

The same unitary ternary analogs of the symmetric series (2.150)–(2.153) for the first and the second circle-like solutions from (2.34) are

$$U_{\text{symm1}}^{8-\text{vertex}}(\alpha,\beta) = \begin{pmatrix} e^{i(\alpha+\beta)} & 0 & 0 & 0 & 0 & 0 & 0 & 0 \\ 0 & 0 & 0 & 0 & \varkappa e^{i(\alpha+\beta)} & 0 & 0 & 0 \\ 0 & 0 & 0 & 0 & 0 & 0 & 0 & e^{2i\beta} \\ 0 & 0 & 0 & e^{i(\alpha+\beta)} & 0 & 0 & 0 & 0 \\ 0 & \pm\varkappa e^{i(\alpha+\beta)} & 0 & 0 & 0 & 0 & 0 & 0 \\ 0 & 0 & 0 & 0 & 0 & e^{i(\alpha+\beta)} & 0 & 0 \\ 0 & 0 & 0 & 0 & 0 & 0 & e^{i(\alpha+\beta)} & 0 \\ 0 & 0 & \pm e^{2i\alpha} & 0 & 0 & 0 & 0 & 0 \end{pmatrix}, \quad (2.257)$$

and

$$U_{\text{symm2}}^{8-\text{vertex}}(\alpha,\beta) = \begin{pmatrix} 0 & 0 & 0 & 0 & 0 & e^{2i\beta} & 0 & 0 \\ 0 & e^{i(\alpha+\beta)} & 0 & 0 & 0 & 0 & 0 & 0 \\ 0 & 0 & e^{i(\alpha+\beta)} & 0 & 0 & 0 & 0 & 0 \\ 0 & 0 & 0 & 0 & 0 & 0 & \varkappa e^{i(\alpha+\beta)} & 0 \\ 0 & 0 & 0 & 0 & e^{i(\alpha+\beta)} & 0 & 0 & 0 \\ \pm e^{2i\alpha} & 0 & 0 & 0 & 0 & 0 & 0 & 0 \\ 0 & 0 & 0 & \pm\varkappa e^{i(\alpha+\beta)} & 0 & 0 & 0 & 0 \\ 0 & 0 & 0 & 0 & 0 & 0 & 0 & e^{i(\alpha+\beta)} \end{pmatrix}, \quad (2.258)$$

respectively.

The invertible 16-vertex star-like solutions (2.206) to the ternary braid equations (2.134) lead to the following two unitary one-parameter ternary braiding quantum gates (cf the binary case (2.245))

$$U_{3-\text{qubits}\pm}^{16-\text{vertex}}(\alpha) = \frac{1}{\sqrt{2}} \begin{pmatrix} e^{3i\alpha} & 0 & 0 & 0 & 0 & 0 & 0 & -1 \\ 0 & e^{3i\alpha} & 0 & 0 & 0 & 0 & \mp e^{2i\alpha} & 0 \\ 0 & 0 & e^{3i\alpha} & 0 & 0 & -e^{2i\alpha} & 0 & 0 \\ 0 & 0 & 0 & e^{3i\alpha} & \mp e^{4i\alpha} & 0 & 0 & 0 \\ 0 & 0 & 0 & \pm e^{2i\alpha} & e^{3i\alpha} & 0 & 0 & 0 \\ 0 & 0 & e^{4i\alpha} & 0 & 0 & e^{3i\alpha} & 0 & 0 \\ 0 & \pm e^{4i\alpha} & 0 & 0 & 0 & 0 & e^{3i\alpha} & 0 \\ e^{6i\alpha} & 0 & 0 & 0 & 0 & 0 & 0 & e^{3i\alpha} \end{pmatrix}. \quad (2.259)$$

The braiding gate (2.259) is a ternary analog of (2.245), and therefore with $\alpha = 0$ it can be treated as a ternary analog of the 8-vertex braiding gate considered in Kauffman and Lomonaco (2004). Note that the solution $U_{3-\text{qubits}+}^{16-\text{vertex}}(0)$ is transpose to the so-called generalized Bell matrix (Rowell *et al* 2010). Comparing (2.184) and (2.259), we observe that the ternary braiding quantum gates (acting on 3 qubits) are those elements of the 16-vertex star semigroup $\mathcal{G}^{\text{star}}_{16\text{vert}}$ (2.192), which satisfy unitarity (2.227).

In the same way, the 32-vertex analog the 8-vertex binary braiding gate of Kauffman and Lomonaco (2004) (now acting on four qubits) is the following constant 4-ary braiding unitary quantum gate

$$U^{32-\text{vertex}}_{4-\text{qubits}} = \frac{1}{\sqrt{2}}\begin{pmatrix} 1&0&0&0&0&0&0&0&0&0&0&0&0&0&0&-1\\ 0&1&0&0&0&0&0&0&0&0&0&0&0&0&-1&0\\ 0&0&1&0&0&0&0&0&0&0&0&0&0&-1&0&0\\ 0&0&0&1&0&0&0&0&0&0&0&0&-1&0&0&0\\ 0&0&0&0&1&0&0&0&0&0&0&-1&0&0&0&0\\ 0&0&0&0&0&1&0&0&0&0&-1&0&0&0&0&0\\ 0&0&0&0&0&0&1&0&0&-1&0&0&0&0&0&0\\ 0&0&0&0&0&0&0&1&-1&0&0&0&0&0&0&0\\ 0&0&0&0&0&0&0&1&1&0&0&0&0&0&0&0\\ 0&0&0&0&0&0&1&0&0&1&0&0&0&0&0&0\\ 0&0&0&0&0&1&0&0&0&0&1&0&0&0&0&0\\ 0&0&0&0&1&0&0&0&0&0&0&1&0&0&0&0\\ 0&0&0&1&0&0&0&0&0&0&0&0&1&0&0&0\\ 0&0&1&0&0&0&0&0&0&0&0&0&0&1&0&0\\ 0&1&0&0&0&0&0&0&0&0&0&0&0&0&1&0\\ 1&0&0&0&0&0&0&0&0&0&0&0&0&0&0&1 \end{pmatrix} \quad (2.260)$$

Thus, in general, the Minkowski star solutions for n-ary braid equations correspond to 2^n-vertex braiding unitary quantum gates as $2^L \times 2^L$ matrices acting on $L = n$ qubits

$$U^{2^L-\text{vertex}}_{L-\text{qubits}} = \frac{1}{\sqrt{2}}\begin{pmatrix} 1&0&0&0&0&-1\\ 0&\ddots&0&0&⋰&0\\ 0&0&1&-1&0&0\\ 0&0&1&1&0&0\\ 0&⋰&0&0&\ddots&0\\ 1&0&0&0&0&1 \end{pmatrix}. \quad (2.261)$$

The braiding gate (2.261) can be treated as a polyadic (n-ary) generalization of the GHZ generator (see, e.g., Rowell *et al* 2010, Ballard and Wu 2011b) acting on $L = n$ qubits.

2.7 Entangling braiding gates

Entangled quantum states are obtained from separable states by acting with special quantum gates on two-qubit states and multi-qubit states (Jaffali and Oeding 2020, 2016). Here we consider the concrete form of braiding gates, which can be entangling or not entangling. There are general considerations on these subjects for the Yang–Baxter maps (Kauffman and Lomonaco 2004, Balakrishnan and Sankaranarayanan 2010, Padmanabhan *et al* 2021) and generalized Yang–Baxter maps (Chen 2012a, Vasquez *et al* 2016, Rowell *et al* 2010, Padmanabhan *et al*

2020b). We present the solutions for the binary and ternary braid maps introduced above, which connect the parameters of the gate and the state.

2.7.1 Entangling binary braiding gates

Let us first examine how the 8-vertex star binary braiding gate $U_s(\alpha, \beta) \equiv U^{8-\text{vert,star}}_{\text{rank}=4}(\alpha, \beta)$ (2.245) acts on the product of one-qubit states concretely. We use the Bloch representation (2.221) to obtain the expression for the transformed concurrence (2.223)

$$C^{(2)}_{s\pm}(U_s(\alpha, \beta) \mid \psi(\theta_1, \gamma_1)\rangle \otimes \mid \psi(\theta_2, \gamma_2)\rangle) \\ = \left| \left(e^{i(\beta+2\gamma_1)} \sin^2 \frac{\theta_1}{2} \pm e^{i\alpha} \cos^2 \frac{\theta_1}{2}\right)\left(e^{i(\beta+2\gamma_2)} \sin^2 \frac{\theta_2}{2} \mp e^{i\alpha} \cos^2 \frac{\theta_2}{2}\right) \right|. \tag{2.262}$$

In general, a braiding gate is *entangling* if the transformed concurrence (2.262) does not vanish, and its roots give the values of the gate parameters $U(\alpha, \beta)$ for which the gate is *not entangling* for a given two-qubit state. In search of the solutions for the transformed concurrence $C^{(2)}_{s\pm} = 0$, we observe that one of the qubits has to be on the Bloch sphere equator $\theta_1 = \frac{\pi}{2}$ (or $\theta_2 = \frac{\pi}{2}$). Only in this case can the first (or second) bracket in (2.262) vanish, and we obtain

$$(1) C^{(2)}_{s+} = 0, \ \text{if } \theta_1 = \frac{\pi}{2} \text{and } \alpha - \beta = 2\gamma_1 - \pi, \ \text{ or } \theta_2 = \frac{\pi}{2} \text{and } \alpha - \beta = 2\gamma_2; \tag{2.263}$$

$$(2) C^{(2)}_{s-} = 0, \ \text{if } \theta_1 = \frac{\pi}{2} \text{and } \alpha - \beta = 2\gamma_1, \ \text{ or } \theta_2 = \frac{\pi}{2} \text{and } \alpha - \beta = 2\gamma_2 - \pi. \tag{2.264}$$

Therefore, the 8-vertex star binary braiding gates (2.245) with the parameters fixed by (2.263)–(2.264) are not entangling.

For the 8-vertex circle binary braiding gate $U_c(\alpha, \beta) \equiv U^{8-\text{vert,circ}}_{\text{rank}=4}(\alpha, \beta)$ (2.252), we obtain

$$C^{(2)}_{c}(U_c(\alpha, \beta) \mid \psi(\theta_1, \gamma_1)\rangle \otimes \mid \psi(\theta_1, \gamma_1)\rangle) \\ = W \left| \left(e^{i(\beta+2\gamma_1)} \sin^2 \frac{\theta_1}{2} - ie^{i\alpha} \cos^2 \frac{\theta_1}{2}\right)\left(e^{i(\beta+2\gamma_2)} \sin^2 \frac{\theta_2}{2} - ie^{i\alpha} \cos^2 \frac{\theta_2}{2}\right) \right|, \tag{2.265}$$

$$W = \frac{(\sqrt{5} - 1)^{\frac{3}{2}}}{\sqrt{2}} = 0.971\,74. \tag{2.266}$$

Analogously to (2.263)–(2.264), the concurrence of the states transformed by the 8-vertex circle binary braiding gate (2.252) can vanish if

$$C^{(2)}_{c} = 0, \ \text{if } \theta_1 = \frac{\pi}{2} \text{ and } \alpha - \beta = 2\gamma_1 - \frac{\pi}{2}, \ \text{ or } \theta_2 = \frac{\pi}{2} \text{ and } \alpha - \beta = 2\gamma_2 - \frac{\pi}{2}. \tag{2.267}$$

Thus, the 8-vertex circle binary braiding gates (2.252) are not entangling if the parameters satisfy (2.267).

In the case of the 4-vertex circle binary braiding gate (2.254), the transformed concurrence vanishes identically, and therefore this gate is not entangling for any values of its parameters.

2.7.2 Entangling ternary braiding gates

Let us consider the tensor product of three qubit pure states $|\psi(\theta_1, \gamma_1)\rangle \otimes |\psi(\theta_2, \gamma_2)\rangle \otimes |\psi(\theta_3, \gamma_3)\rangle$ (in the Bloch representation (2.221)), which obviously has zero concurrence $C^{(3)}$ (2.225) because of the vanishing of the hyperdeterminant (2.226). After transforming by the 16-vertex star ternary braiding gates $U_{16}(\alpha) \equiv U_{3-\text{qubits}}^{16-\text{vertex}}(\alpha)$ (2.259), the concurrence becomes

$$C_{16\pm}^{(3)}(U_{16}(\alpha)\,|\psi(\theta_1, \gamma_1)\rangle \otimes |\psi(\theta_2, \gamma_2)\rangle \otimes |\psi(\theta_3, \gamma_3)\rangle)$$
$$= \frac{1}{64}\,|\,e^{2i\alpha} \pm e^{2i\gamma_1} + (e^{2i\alpha} \mp e^{2i\gamma_1})\cos\theta_1\,|^2\,|\,e^{2i\alpha} - e^{2i\gamma_2} + (e^{2i\alpha} + e^{2i\gamma_2})\cos\theta_2\,|^2 \tag{2.268}$$
$$|\,e^{2i\alpha} \mp e^{2i\gamma_3} + (e^{2i\alpha} \pm e^{2i\gamma_3})\cos\theta_3\,|^2.$$

We observe that the ternary concurrence (2.268) vanishes if any of the brackets is equal to zero. Because the domain of all angles is $\mathbb{R}$, we only have solutions for fixed discrete $\theta_k = \pi, -\pi, \pi/2$, $k = 1, 2, 3$, which means that on the Bloch sphere the quantum states should be on the equator (as in the binary case), or additionally at the poles. In this case, $e^{i\alpha} = \pm e^{i\gamma_k}$, and

$$\alpha = \begin{cases} \gamma_k \\ \gamma_k + \pi \end{cases}, \quad k = 1, 2, 3. \tag{2.269}$$

Thus, for a fixed three-qubit product state one (or more) of which is at a pole or the equator of the Bloch sphere, those ternary braiding gates $U_{16}(\alpha)$ satisfying the conditions (2.269) are not entangling $C_{16\pm}^{(3)} = 0$, whereas in other cases they are entangling $C_{16\pm}^{(3)} \neq 0$.

By analogy, a similar action of the 8-vertex bisymmetric (star-like) ternary braiding gates $U_{8b1,\,2}(\alpha, \beta) \equiv U_{\text{bisymm}1,\,2}^{8-\text{vertex}}(\alpha, \beta)$ (2.255)–(2.256) gives

$$C_{8b1}^{(3)}(U_{8b1}(\alpha, \beta)\,|\psi(\theta_1, \gamma_1)\rangle \otimes |\psi(\theta_2, \gamma_2)\rangle \otimes |\psi(\theta_3, \gamma_3)\rangle)$$
$$= \left|\,\sin^2\theta_1 \sin^2\theta_3 \left(e^{2i(\beta+\gamma_2)}\sin^2\frac{\theta_2}{2} - e^{2i\alpha}\cos^2\frac{\theta_2}{2}\right)^2\,\right|, \tag{2.270}$$

$$C_{8b2}^{(3)}(U_{8b2}(\alpha, \beta)\,|\psi(\theta_1, \gamma_1)\rangle \otimes |\psi(\theta_2, \gamma_2)\rangle \otimes |\psi(\theta_3, \gamma_3)\rangle)$$
$$= \left|\,\sin^2\theta_1 \sin^2\theta_3 \left(e^{2i(\alpha+\gamma_2)}\sin^2\frac{\theta_2}{2} - e^{2i\beta}\cos^2\frac{\theta_2}{2}\right)^2\,\right|. \tag{2.271}$$

Their solutions coincide with the binary case (2.263)–(2.264) applied to the middle qubit $|\psi(\theta_2, \gamma_2)\rangle$ and $\gamma_2 \to 2\gamma_2$.

The action of the 8-vertex symmetric (circle-like) ternary braiding gates $U_{8s}(\alpha, \beta) \equiv U^{8-\text{vertex}}_{\text{symm1, 2}}(\alpha, \beta)$ (2.257)–(2.258) leads to the transformed concurrence

$$C^{(3)}_{8s}(U_{8s}(\alpha, \beta) \mid \psi(\theta_1, \gamma_1)\rangle \otimes \mid \psi(\theta_2, \gamma_2)\rangle \otimes \mid \psi(\theta_3, \gamma_3)\rangle) \\ = \left| \sin^2 \theta_2 \left(e^{i(\beta+2\gamma_1)} \sin^2 \frac{\theta_1}{2} - e^{i\alpha} \cos^2 \frac{\theta_1}{2} \right) \left(e^{i(\beta+2\gamma_3)} \sin^2 \frac{\theta_3}{2} - e^{i\alpha} \cos^2 \frac{\theta_3}{2} \right) \right|. \quad (2.272)$$

The conditions for this to vanish, i.e., when the gate $U_{8s}(\alpha, \beta)$ becomes not entangling, coincide with those for the binary case (2.263)–(2.264), applied here to the first or the third qubit.

Thus we have shown that the braiding binary and ternary quantum gates can be either entangling or not entangling, depending on how their parameters are related to the concrete quantum state on which they act.

The constructions presented here (Duplij and Vogl 2021) could be used, e.g., in the entanglement-free protocols (de Burgh and Bartlett 2005, Rehman and Shin 2021) and some experiments (Almeida *et al* 2014, Higgins *et al* 2007). This can also allow us to build quantum networks without any entangling at all *non-entangling networks*, when the next gate depends upon the previous state in such a way that at each step there is no entangling because the separable, but different, final state is received from a separable initial state.

References

Aasen D *et al* 2016 Milestones toward Majorana-based quantum computing *Phys. Rev.* X **6** 031016

Alagic G, Bapat A and Jordan S 2014 Classical simulation of Yang-Baxter gates *9th Conf. Theory of Quantum Computation, Communication and Cryptography vol 27 of LIPIcs. Leibniz Int. Proc. Inform.* ed S T Flammia and A W Harrow (Dagstuhl: Schloss Dagstuhl. Leibniz-Zent. Inform., Wadern) 161–75 pp

Alagic G, Jarret M and Jordan S P 2016 Yang-Baxter operators need quantum entanglement to distinguish knots *J. Phys. A. Math. Theor.* **49** 12

Alagic G, Jeffery S and Jordan S 2014 Circuit obfuscation using braids *9th Conf. Theory of Quantum Computation, Communication and Cryptography vol 27 of LIPIcs. Leibniz Int. Proc. Inform.* (Schloss Dagstuhl, Wadern: Leibniz-Zent. Inform.) 141–60 pp

Almeida M P, Gu M, Fedrizzi A, Broome M A, Ralph T C and White A G 2014 Entanglement-free certification of entangling gates *Phys. Rev.* **A89** 042323

Aravind P K 1997 Borromean entanglement of the GHZ state *Potentiality, Entanglement and Passion-at-a-Distance vol 194 of Boston Stud. Philos. Sci.* ed R S Cohen, M Horne and J Stachel (Boston: Kluwer Acad. Publ.) 53–9 pp

Asaulko K S and Korablev P G 2019 On a Yang-Baxter operator and the corresponding knots invariant *Chelyabinsk Phys. Math. J.* **4** 255–64

Aziz R K and Muchtadi-Alamsyah I 2015 Braid group representation on quantum computation *AIP Conference Proceedings: The 5Th Int. Conf. Mathematics And Natural Sciences* vol 1677 (New York: AIP Publishing) 030011 pp

Balakrishnan S and Sankaranarayanan R 2010 Entangling power and local invariants of two-qubit gates *Phys. Rev.* A **82** 034301

Ballard A D and Wu Y-S 2011a Cartan decomposition and entangling power of braiding quantum gates *Cross Disciplinary Advances in Quantum Computing* (Providence: Amer. Math. Soc.) 1–15 pp

Ballard A D and Wu Y-S 2011b Cartan decomposition and entangling power of braiding quantum gates *Cross Disciplinary Advances in Quantum Computing vol 536 of Contemp. Math.* ed K Mahdavi, D Koslover and L L Brown (Amer. Math. Soc.: Providence) 1–15 pp

Behrends J and Béri B 2022 Sachdev-Ye-Kitaev circuits for braiding and charging majorana zero modes *Phys. Rev. Lett.* **128** 106805

Bombin H and Martin-Delgado M A 2007 Topological computation without braiding *Phys. Rev. Lett.* **98** 160502

Bonderson P, Freedman M and Nayak C 2008 Measurement-only topological quantum computation *Phys. Rev. Lett.* **101** 4

Bonesteel N E, Hormozi L, Zikos G and Simon S H 2005 Braid topologies for quantum computation *Phys. Rev. Lett.* **95** 140503

Bruzda W, Friedland S and Zyczkowski K 2019 *Tensor Rank and Entanglement of Pure Quantum States* 36 Jagiellonian University Krakow 36 p (arxiv:1912.06854)

Bukhshtaber V M 1998 The Yang-Baxter transformation *Russ. Math. Surv.* **53** 1343–4

Bullock S S 2004 Note on the Khaneja Glaser decomposition *Quantum Inf. Comp.* **4** 396–400

Bullock S S and Brennen G K 2004 Canonical decompositions of n-qubit quantum computations and concurrence *J. Math. Phys.* **45** 2447–67

Chen R S 2012a Generalized Yang-Baxter equations and braiding quantum gates *J. Knot Theory Ramifications* **21** 18

Chen R S 2012b Generalized Yang-Baxter equations and braiding quantum gates *J. Knot Theor. Ramif* **21** 1250087

Chong B, Keiter H and Stolze J 2005 *Multiparticle entanglement and ranks of density matrices* 4

Conlon A and Slingerland J K 2022 Compatibility of braiding and fusion on wire networks *Phys. Rev.* B **108** 035150

Cui S X, Hong S-M and Wang Z 2015 universal quantum computation with weakly integral anyons *Quantum Inf. Process.* **14** 2687–727

Dai L, Li D, Tu Q and Wang J 2022 Yarn density measurement for 3-D braided composite preforms based on rotation object detection *IEEE Trans. Instrum. Meas.* **71** 1–11

Darunkar B and Verma P 2014 The braided single-stage protocol for quantum secure communication *Quantum Information and Computation XII* vol 9123 ed E Donkor, A R Pirich, H E Brandt, M R Frey, S J Lomonaco and J M Myers (Bellingham, Washington USA: SPIE) 912308

de Burgh M and Bartlett S D 2005 Quantum methods for clock synchronization: Beating the standard quantum limit without entanglement *Phys. Rev.* **A72** 9

Dörnte W 1929 Unterschungen über einen verallgemeinerten Gruppenbegriff *Math. Z.* **29** 1–19

Drinfeld V G 1989 Quasi-Hopf algebras and Knizhnik-Zamolodchikov equations *Problems of Modern Quantum Field Theory* ed A A Belavin, A V Klimyk and A B Zamolodchikov (Heidelberg: Springer) pp 1–13

Drinfeld V G 1992 On some unsolved problems in quantum group theory *Quantum Groups* ed P P Kulish (Berlin-Heidelberg: Springer) pp 1–8

Duplij S 2001 Ternary Hopf algebras *Symmetry in Nonlinear Mathematical Physics* ed A A Nikitin and V M Boyko (Kiev: Institute of Mathematics) pp 25–34

Duplij S 2018 Polyadic algebraic structures and their representations *Exotic Algebraic and Geometric Structures in Theoretical Physics* ed S Duplij (New York: Nova Publishers) pp 251–308 (arXiv:math.RT/1308.4060)

Duplij S 2021a Higher braid groups and regular semigroups from polyadic-binary correspondence *Mathematics* **9** 972

Duplij S 2021b Polyadic Hopf algebras and quantum groups *East European J. Phys.* **2** 5–50

Duplij S 2022 *Polyadic Algebraic Structures* (Bristol: IOP Publishing)

Duplij S and Vogl R 2021 Polyadic braid operators and higher braiding gates *Universe* **7** 301

Dye H A 2003 Unitary solutions to the Yang-Baxter equation in dimension four *Quant. Inf. Processing* **2** 117–50

Dye H A and Kauffman L H 2011 Anyonic topological quantum computation and the virtual braid group *J. Knot Theory Ramifications* **20** 91–102

Epping M, Kampermann H, Macchiavello C and Bruß C 2017 Multi-partite entanglement can speed up quantum key distribution in networks *New J. Phys.* **19** 093012

Ezawa M 2023 *Topological Quantum Gates and Topological Entangled States by Braiding Majorana Fermions* 15 preprint Univ. Tokyo Tokyo 15 p (2304.06260)

Fan Z and de Garis H 2010 Braid matrices and quantum gates for ising anyons topological quantum computation *Eur. Phys. J.* B **74** 419–27

Fowler A G, Mariantoni M, Martinis J M and Cleland A N 2012 Surface codes: Towards practical large-scale quantum computation *Phys. Rev.* A **86** 032324

Fraïsse J M E and Braun D 2016 Quantum channel-estimation with particle loss: GHZ versus W states *Quantum Measurements and Quantum Metrology* **3** 53–67

Freedman M H, Kitaev A, Larsen M J and Wang Z 2003 Topological quantum computation *Bull. Amer. Math. Soc. (N.S.)* **40** 31–8

Garnerone S, Marzuoli A and Rasetti M 2007 Quantum automata, braid group and link polynomials *Quantum Inform. Compu.* **7** 479–503

Génetay Johansen E and Simula T 2021 Prime number factorization using a spinor bose-einstein condensate-inspired topological quantum computer *Quantum Inf. Process.* **21** 1–19

Hanks M, Estarellas M P, Munro W J and Nemoto K 2020 Effective compression of quantum braided circuits aided by ZX-calculus *Phys. Rev.* X **10** 041030

Harle N, Shtanko O and Movassagh R 2023 Observing and braiding topological majorana modes on programmable quantum simulators *Nat. Commun.* **14** 2286

Helgason S 1962 *Differential Geometry and Symmetric Spaces* (Academic Press: New York–London)

Hietarinta J 1993 Solving the two-dimensional constant quantum Yang-Baxter equation *J. Math. Phys.* **34** 1725–56

Higgins B L, Berry D W, Bartlett S D, Wiseman H M and Pryde G J 2007 Entanglement-free Heisenberg-limited phase estimation *Nature* **450** 393–6

Ho C L and Deguchi T 2017 Multi-qudit states generated by unitary braid quantum gates based on Temperley-Lieb algebra *Europhys. Lett.* **118** 40001

Ho C-L, Solomon A I and Oh C-H 2010 Quantum entanglement, unitary braid representation and Temperley-Lieb algebra *Europhys. Lett.* **92** 30002

Hoffman S, Schrade C, Klinovaja J and Loss D 2016 universal quantum computation with hybrid spin-Majorana qubits *Phys. Rev.* B **94** 045316

Horsman C, Fowler A G, Devitt S and Meter R V 2012 Surface code quantum computing by lattice surgery *New J. Phys.* **14** 27 Id/No 123 011

Hu T, Wang G, Sun C, Zhou C, Wang Q and Xue K 2009 Method of constructing braid group representation and entanglement in a 9×9 Yang-Baxter system *Rev. Math. Phys.* **21** 1081–90

Huang H-L *et al* 2021 Emulating quantum teleportation of a majorana zero mode qubit *Phys. Rev. Lett.* **126** 090502

Hutter A and Loss D 2016 Quantum computing with parafermions *Phys. Rev.* B **93** 125105

Jaffali H and Oeding L 2020 Learning algebraic models of quantum entanglement *Quantum Inf. Process.* **19** 279

Jaworowski B and Hawrylak P 2019 Quantum bits with macroscopic topologically protected states in semiconductor devices *Appl. Sci.* **9** 474

Johansen E G and Simula T 2021 Fibonacci anyons versus majorana fermions *PRX Quantum* **2** 010334

Kassel C 1995 *Quantum Groups* (New York: Springer)

Kauffman L H 2017 Braiding and Majorana fermions *J. Knot Theory Ramifications* **26** 21

Kauffman L H and Lomonaco S J 2002 Quantum entanglement and topological entanglement *New J. Phys.* **4** 73.1–18

Kauffman L H and Lomonaco S J 2004 Braiding operators are universal quantum gates *New J. Phys.* **6** 134–9

Kauffman L H and Lomonaco S J 2018 Braiding, Majorana fermions, Fibonacci particles and topological quantum computing *Quantum Inf. Process.* **17** 81

Kauffman L H and Mehrotra E 2019 Topological aspects of quantum entanglement *Quantum Inf. Process.* **18** 36

Kaye P, Laflamme R and Mosca M 2007 *An introduction to Quantum Computing* (Oxford: Oxford Univ. Press)

Khaneja N and Glaser S J 2001 Cartan decomposition of $SU(2n)$ and control of spin systems *Chem. Phys.* **267** 11–23

Kitaev A and Wang Z 2012 Solutions to generalized Yang-Baxter equations via ribbon fusion categories *Proceedings of the Freedman Fest vol 18 of Geom. Topol. Monogr.* (Coventry: Geom. Topol. Publ.) pp 191–7

Kolganov N, Mironov S and Morozov A 2021 *Large k* Topological Quantum Computer preprint MIPT Moscow 14 p (arxiv:2105.03980)

Kolganov N and Morozov A 2020 Quantum R-matrices as universal qubit gates *JETP Letters* **111** 519–24

Kraus B and Cirac J I 2001 Optimal creation of entanglement using a two-qubit gate *Phys. Rev.* **A63** 8

Lambe L A and Radford D E 1997 *Introduction to the Quantum Yang-Baxter equation and Quantum Groups: An Algebraic Approach* (Dordrecht: Kluwer)

Lusztig G 2010 Braid group relations *Introduction to Quantum Groups* (Boston: Birkhuser) pp 304–17

Ma X, Reichhardt C J O and Reichhardt C 2020 Braiding majorana fermions and creating quantum logic gates with vortices on a periodic pinning structure *Phys. Rev.* B **101** 024514

Majid S 2022 Quantum and braided ZX calculus *J. Phys. A: Math. Theor.* **55** 34

Martin I and Agarwal K 2020 Double braiding majoranas for quantum computing and hamiltonian engineering *PRX Quantum* **1** 020324

Matsumoto D K and Shibukawa Y 2015 Quantum Yang-Baxter equation, braided semigroups and dynamical Yang-Baxter maps *Tokyo J. Math.* **38** 227–37

Melnikov D, Mironov A, Mironov S, Morozov A and Morozov A 2018 Towards topological quantum computer *Nucl. Phys.* **B926** 491–508

Melnikov D, Mironov A, Mironov S, Morozov A and Morozov A 2019 From topological to quantum entanglement *J. High Energy Phys.* **5** 11

Mi S, Pikulin D I, Wimmer M and Beenakker C W J 2013 Proposal for the detection and braiding of Majorana fermions in a quantum spin Hall insulator *Phys. Rev.* B **87** 241405

Mong R S K 2014 universal topological quantum computation from a superconductor-abelian quantum hall heterostructure *Phys. Rev.* X **4** 011036

Nashed M Z 1976 *Generalized Inverses and Applications* (New York: Academic)

Nayak C, Simon S H, Stern A, Freedman M and Das Sarma S 2008 Non-abelian anyons and topological quantum computation *Rev. Mod. Phys.* **80** 1083–159

Nayak C, Simon S H, Stern A, Freedman M and Das Sarma S 2008 Non-abelian anyons and topological quantum computation *Rev. Mod. Phys.* **80** 1083–159

Neven A, Martin J and Bastin T 2018 Entanglement robustness against particle loss in multiqubit systems *Phys. Rev.* **A98** 062335

Nielsen M A and Chuang I L 2000 *Quantum Computation and Quantum Information* (Cambridge: Cambridge University Press)

Ootsuka T and Sakuma K 2008 Braid group and topological quantum computing *Mathematical Aspects of Quantum Computing* (Hackensack: World Scientific) pp 55–89

Padmanabhan P, Sugino F and Trancanelli D 2020a Braiding quantum gates from partition algebras *Quantum* **4** 311

Padmanabhan P, Sugino F and Trancanelli D 2020b Generating W states with braiding operators *Quantum Inf. Comput.* **13 & 14** 1154–62

Padmanabhan P, Sugino F and Trancanelli D 2021 Local invariants of braiding quantum gates—associated link polynomials and entangling power *J. Phys. A: Math. Theor.* **54** 135301

Park A J, McKay E, Lu D and Laflamme R 2016 Simulation of anyonic statistics and its topological path independence using a seven-qubit quantum simulator *New J. Phys.* **18** 15

Rehman J and Shin H 2021 Entanglement-free parameter estimation of generalized Pauli channels *Quantum* **5** 490

Rowell E 2011 A quaternionic braid representation (after goldschmidt and jones) *Quantum Topol.* **2** 173–82

Rowell E C and Wang Z 2018 Mathematics of topological quantum computing *Bull. Amer. Math. Soc. (N. S.)* **55** 183–238

Rowell E C, Zhang Y, Wu Y-S and Ge M-L 2010 Extraspecial two-groups, generalized Yang-Baxter equations and braiding quantum gates *Quantum Inf. Comput.* **10** 685–702

Sau J D, Tewari S and Das Sarma S 2010 universal quantum computation in a semiconductor quantum wire network *Phys. Rev.* A **82** 052322

Stenger J P T, Bronn N T, Egger D J and Pekker D 2020 Simulating the dynamics of braiding of majorana zero modes using an IBM quantum computer *Phys. Rev. Research* **3** 033171

Sudbery A 2001 On local invariants of pure three-qubit states *J. Phys. A: Math. Gen.* **34** 643–52

Turaev V G 1988 The Yang-Baxter equation and invariants of links *Invent. Math.* **92** 527–53

Vartiainen J J, Niskanen A O, Nakahara M and Salomaa M M 2004 Acceleration of quantum algorithms using three-qubit gates *Inter. J. Quantum Inf.* **2** 1–10

Vasquez J F, Wang Z and Wong H M 2016 Qubit representations of the braid groups from generalized Yang-Baxter matrices *Quantum Inf. Process.* **15** 3035–42

Veselov A P 2003 Yang-Baxter maps and integrable dynamics *Phys. Lett.* **A314** 214–21

Walter M, Gross D and Eisert J 2016 Multipartite entanglement *Quantum Information* ed D Bruss and G Leuchs (New York: Wiley) pp 293–330

Wasa K, Nishio S, Suetsugu K, Hanks M, Stephens A, Yokoi Y and Nemoto K 2023 Hardness of braided quantum circuit optimization in the surface code *IEEE Trans. Quantum Eng.* **4** 1–7

Williams C P and Clearwater S H 1998 *Explorations in Quantum Computing* (New York-Berlin-Heidelberg: Springer)

Wu L-H, Liang Q-F and Hu X 2014 New scheme for braiding Majorana fermions *Sci. Technol. Adv. Mater.* **15** 064402

Yu G C, Wang X F and Xue K 1992 Braid group representation and solutions of the quantum Yang-Baxter equation associated with 10-dimensional representation of $SU(4)$ *Commun. Theor. Phys.* **18** 171–8

Zangi S M and Qiao C-F 2021 Robustness of $2 \times N \times M$ entangled states against qubit loss *Phys. Lett.* **400** 6

Zhang H *et al* 2018 Quantized Majorana conductance *Nature* **556** 74–9

Zhou Y-F, Hou Z and Sun Q-F 2018 Non-abelian braiding of chiral Majorana fermions by quantum dots *Phys. Rev.* B **99** 195137

Zhu G, Lavasani A and Barkeshli M 2020 universal logical gates on topologically encoded qubits via constant-depth unitary circuits *Phys. Rev. Lett.* **125** 050502

Zhu P and Wen Q 2011 Affine braid groups: a better platform than braid groups for cryptology? *Applicable Algebra in Engineering, Communication and Computing* vol 22 (Berlin: Springer) pp 375–91

IOP Publishing

Steven Duplij and Raimund Vogl

Chapter 3

Supersymmetry and quantum computing

It is well-known that quantum computation is based on the algebraic structure of its constituents, qubits and qudits, living in some Hilbert space. Therefore, possible improvements could be connected with some special generalizations of the Hilbert space. A promising direction is supersymmetric generalization of the ordinary Hilbert space (De Witt 1992, Constantinescu 2002) and consideration of various super analogs of quantum states in it, with simultaneous passing from corresponding groups to supergroups.

3.1 Superspaces and supermatrices

Let us consider the main ideas in the supersymmetrization of qubits (Borsten *et al* 2010, 2014, 2015). The principal statement changes the Hilbert space to super Hilbert space (in sense of Rudolph 2000) and considers quantum states as (even) supervectors in the latter, i.e., taking values in the corresponding Grassmann algebra (or some more general supercommutative superalgebra). In this approach, the inner product and probabilities contain Grassmann algebra parts. In the same way, the bra/ket formalism of quantum mechanics (Dirac 1939, van Eijndhoven and de Graaf 1985, Gieres 2000) transforms to super-bra/super-ket formalism with additional parity rules. Here we will point out the foremost relations and statements concisely (only needed), while we refer to the details and further notations to the standard supermathematics sources (Berezin 1987, Leites 1980, De Witt 1992). To clarify the structure of variables, we present some formulas in two columns: ordinary (left) and supersymmetric (right) cases, and moreover we use different notations for them (the latter will be marked in bold).

Let $\Lambda_N(\mathbb{C})$ be a complex Grassmann algebra having N anticommuting generators θ_i. The nilpotence of θs (which follows from their anticommutativity) leads to its finiteness (with dimension 2^N) and to the decompositions of any element $z \in \Lambda_N(\mathbb{C})$ (informally)

doi:10.1088/978-0-7503-5281-9ch3

$$z = \overbrace{z_{\text{body}}}^{\text{no } \theta's} + \overbrace{z_{\text{soul}}}^{\text{with } \theta's} = \overbrace{z_{\text{invert}}}^{\text{no } \theta's \text{ and no } 0} + \overbrace{z_{\text{noninvert}}}^{\text{with } \theta's \text{ and } 0} = \overbrace{z_{\text{even}}}^{\text{even } \theta's} + \overbrace{z_{\text{odd}}}^{\text{odd } \theta's}, \tag{3.1}$$

where $z_{\text{body}} \in \mathbb{C}$, $z_{\text{invert}} \in \mathbb{C}\backslash 0$, $z_{\text{soul}} \in \Lambda_N(\mathbb{C})\backslash\mathbb{C}$, $z_{\text{noninvert}} \in \Lambda_N(\mathbb{C})\backslash\mathbb{C} \cup \{0\}$. The last decomposition allows us to introduce the degree by $\deg z_{\text{even}} = \bar{0}$ and $\deg z_{\text{odd}} = \bar{1}$, and elements with the fixed degree are homogeneous. Obviously, that for homogeneous elements $\deg yz = \deg\ y + \deg z (\text{mod } 2)$. Another name of deg is parity (or grade), in special cases they are fine different (Bernstein *et al* 2013) but in the superqubit context all of them are interchangable. Thus, the mapping $\deg\colon \Lambda_N(\mathbb{C}) \to \mathbb{Z}_2 = \{\bar{0}, \bar{1}\}$, gives the direct sum decomposition of the Grassmann algebra $\Lambda_N(\mathbb{C}) = \Lambda_N^{(\text{even})}(\mathbb{C}) \oplus \Lambda_N^{(\text{odd})}(\mathbb{C}) = \Lambda_N^{(\bar{0})}(\mathbb{C}) \oplus \Lambda_N^{(\bar{1})}(\mathbb{C})$, which means that $\Lambda_N(\mathbb{C})$ is the simplest example of $\mathbb{Z}_2$-graded algebra. The analog of the ordinary commutator for $\Lambda_N(\mathbb{C})$ is the supercommutator

$$[y, z] = yz - zy,\ \ y, z \in \mathbb{C}, \overset{\text{susy}}{\Longrightarrow} [y, z]_{\deg} = yz - (-1)^{\deg y \deg z} zy,\ \ y, z \in \Lambda_N(\mathbb{C}). \tag{3.2}$$

If $[y, z]_{\deg} = 0$ for all elements of a superalgebra, then it is supercommutative, which is indeed the case of the Grassmann algebra $\Lambda_N(\mathbb{C})$. The same rule will be implied for all of the other $\mathbb{Z}_2$-graded homogeneous variables.

The ordinary involution $*$ and the grade involution $\sharp$ (superstar or super-involution (Bernstein *et al* 2013, Borsten *et al* 2010)) can be defined on $\Lambda_N(\mathbb{C})$ as follows

$$(xy)^* = \bar{x}y,\ \ (yz)^* = z^* y^*,\ \ (y)^{**} = y, \tag{3.3}$$

$$(xy)^\sharp = \bar{x}y^\sharp,\ \ (yz)^\sharp = y^\sharp z^\sharp,\ \ (y)^{\sharp\sharp} = (-1)^{\deg y} y,\quad x \in \mathbb{C},\ \ y, z \in \Lambda_N(\mathbb{C}), \tag{3.4}$$

such that $z^* \in \Lambda_N^{(\deg z)}(\mathbb{C})$, $z^\sharp \in \Lambda_N^{(\deg z)}(\mathbb{C})$.

The superqubits live in a finite-dimensional $\mathbb{Z}_2$-graded linear vector space (or superspace) $\mathscr{V}$ over $\mathbb{C}$ (or any other field $\mathbb{K}$), which has the same decomposition on the even and odd parts as the Grassmann algebra above $\mathscr{V} = \mathcal{V}^{(\bar{0})} \oplus \mathcal{V}^{(\bar{1})}$. If the dimensions of the component spaces $\dim \mathcal{V}^{(\bar{0})} = p$ and $\dim \mathcal{V}^{(\bar{1})} = q$, then we denote the $\mathbb{Z}_2$-graded vector space $\mathscr{V} = \mathbb{C}^{p|q}$, and its graded dimension $\dim \mathbb{C}^{p|q} = p + q$. The $\mathbb{Z}_2$-graded direct product of superspaces $\hat{\otimes}$ (which is used for superqubit constructions) is crucially different from the ordinary direct product of spaces $\otimes$ (exploited for qubits). Indeed, we have the ordinary direct product

$$\mathscr{V} \otimes \mathscr{W} = \mathcal{V}^{(\bar{0})} \otimes \mathcal{W}^{(\bar{0})} + \mathcal{V}^{(\bar{1})} \otimes \mathcal{W}^{(\bar{1})} + \mathcal{V}^{(\bar{1})} \otimes \mathcal{W}^{(\bar{0})} + \mathcal{V}^{(\bar{0})} \otimes \mathcal{W}^{(\bar{1})}, \tag{3.5}$$

which does not allow us to introduce the $\mathbb{Z}_2$-graded structure without additional assumptions.

Innovation 3.1. *Only the definition of a new operation, the $\mathbb{Z}_2$-graded direct product $\hat{\otimes}$, gives the consistent superspace structure of product (by endowing two last terms in (3.5) with odd grading of the product)*

$$(\mathscr{V} \hat{\otimes} \mathscr{W})^{(\mathsf{k})} = \bigoplus_{\mathsf{k}=\mathsf{r} \boxplus \mathsf{s}} \mathcal{V}^{(\mathsf{r})} \otimes \mathcal{W}^{(\mathsf{s})}, \qquad \mathsf{k}, \mathsf{r}, \mathsf{s} \in \mathbb{Z}_2, \quad \mathsf{r} \boxplus \mathsf{s} \equiv \mathsf{r} + \mathsf{s} (\mathrm{mod}\, 2), \tag{3.6}$$

or simply

$$(\mathscr{V} \hat{\otimes} \mathscr{W})^{(\bar{0})} = \mathcal{V}^{(\bar{0})} \otimes \mathcal{W}^{(\bar{0})} + \mathcal{V}^{(\bar{1})} \otimes \mathcal{W}^{(\bar{1})}, \tag{3.7}$$

$$(\mathscr{V} \hat{\otimes} \mathscr{W})^{(\bar{1})} = \mathcal{V}^{(\bar{1})} \otimes \mathcal{W}^{(\bar{0})} + \mathcal{V}^{(\bar{0})} \otimes \mathcal{W}^{(\bar{1})}. \tag{3.8}$$

Usually, the (different) operations $\hat{\otimes}$ *and* $\otimes$ *are denoted by the same symbol, but they should be used with care and taking account the actual distinction of (3.5) and (3.6).*

In the consideration of mappings between superspaces and trying to introduce $\mathbb{Z}_2$-graded structure for them, we also note some peculiarities (important for superqubit constructions). Indeed, the set of homomorphisms $\{\mathbf{T}\}$ from superspace $\mathscr{V}$ to superspace $\mathscr{W}$ is defined in the standard way

$$\mathrm{Hom}(\mathscr{V}, \mathscr{W}) = \{\mathbf{T} | \mathbf{T}\, \mathscr{V} \subset \mathscr{W}\}. \tag{3.9}$$

We could assume that the $\mathbb{Z}_2$-graded structure is analogous to (3.6)

$$\mathrm{Hom}^{(\mathsf{k})}(\mathscr{V}, \mathscr{W}) = \{\mathbf{T} \in \mathrm{Hom}(\mathscr{V}, \mathscr{W}) | \mathbf{T}\, \mathcal{V}^{(\mathsf{r})} \subset \mathcal{W}^{(\mathsf{r} \boxplus \mathsf{k})}\}, \qquad \mathsf{k}, \mathsf{r} \in \mathbb{Z}_2. \tag{3.10}$$

Innovation 3.2. *Only even mappings* $\mathbf{T}^{(\bar{0})} = \mathrm{Hom}^{(\bar{0})}(\mathscr{V}, \mathscr{W})$ *(*$\deg \mathbf{T} = \bar{0} \in \mathbb{Z}_2$*) are homomorphisms. Only odd mappings* $\mathbf{T}^{(\bar{1})}$ *(*$\deg \mathbf{T} = \bar{1} \in \mathbb{Z}_2$*) are not morphisms at all because they cannot be composed:* $\mathbf{T}^{(\bar{1})} \circ \mathbf{T}^{(\bar{1})}$ *is not odd, but is even mapping.*

The same observation can be made for the linear operators in a $\mathbb{Z}_2$-graded linear vector space $\mathbb{C}^{p|q}$ given by (super)matrices. In the standard basis, a linear operator $\mathbf{T} \in \mathrm{End}(\mathbb{C}^{p|q})$ can be represented by the square block $(p+q) \times (p+q)$ supermatrix over $\Lambda_N(\mathbb{C})$ (Berezin 1987, Leites 1980) (other representations are also possible Bernstein *et al* 2013)

$$\mathrm{M} = \begin{pmatrix} A_{p\times p} & B_{p\times q} \\ C_{q\times p} & D_{q\times q} \end{pmatrix} \in \mathrm{Mat}(p|q, \Lambda_N(\mathbb{C})), \tag{3.11}$$

where the even (ordinary) matrices $A_{p\times p}$, $D_{q\times q}$ are over $\Lambda_N^{(\mathrm{even})}(\mathbb{C}) = \Lambda_N^{(\bar{0})}(\mathbb{C})$, and the odd (ordinary) matrices $B_{p\times q}$, $C_{q\times p}$ are over $\Lambda_N^{(\mathrm{odd})}(\mathbb{C}) = \Lambda_N^{(\bar{1})}(\mathbb{C})$. This full supermatrix has the total parity (degree) $\deg \mathrm{M} = 0$ and

$$\mathrm{M}_{\deg M=0} = \mathrm{M}_{\mathrm{body}} + \mathrm{M}_{\mathrm{soul}} = \begin{pmatrix} \left(A_{p\times p}\right)_{\mathrm{body}} & 0_{p\times q} \\ 0_{q\times p} & \left(D_{q\times q}\right)_{\mathrm{body}} \end{pmatrix} + \mathrm{M}_{\mathrm{soul}}. \tag{3.12}$$

If oppositely, $A_{p\times p}$, $D_{q\times q}$ are (ordinary) matrices over $\Lambda_N^{(\bar{1})}(\mathbb{C})$, and $B_{p\times q}$, $C_{q\times p}$ are (ordinary) matrices over $\Lambda_N^{(\bar{0})}(\mathbb{C})$, then deg M = 1, and

$$\mathrm{M}_{\deg \mathrm{M}=1} = \mathrm{M}_{\mathrm{body}} + \mathrm{M}_{\mathrm{soul}} = \begin{pmatrix} 0_{p\times p} & \left(B_{p\times q}\right)_{\mathrm{body}} \\ \left(C_{q\times p}\right)_{\mathrm{body}} & 0_{q\times q} \end{pmatrix} + \mathrm{M}_{\mathrm{soul}}. \tag{3.13}$$

Innovation 3.3. The supermatrices with deg M = 1 are not morphisms of $\mathbb{C}^{p|q}$ because their product gives the first ones having deg M = 0, and therefore

$$\{\mathrm{M}\}_{\deg \mathrm{M}=1} \notin \mathrm{End}(\mathbb{C}^{p|q}). \tag{3.14}$$

After the decomposition of the matrices (3.11) with deg M = 0 (reminding (3.12), (3.13))

$$\mathrm{M} = \mathrm{M}^{(\mathrm{even})} + \mathrm{M}^{(\mathrm{odd})} = \mathrm{M}^{(0)} + \mathrm{M}^{(1)}, \tag{3.15}$$

$$\mathrm{M}^{(0)} = \begin{pmatrix} A_{p\times p} & 0_{p\times q} \\ 0_{q\times p} & D_{q\times q} \end{pmatrix}, \tag{3.16}$$

$$\mathrm{M}^{(1)} = \begin{pmatrix} 0_{p\times p} & B_{p\times q} \\ C_{q\times p} & 0_{q\times q} \end{pmatrix}, \tag{3.17}$$

we observe that $\mathrm{M}^{(\bar{0})}\mathrm{M}'^{(\bar{0})} = \mathrm{M}''^{(\bar{0})}$, and therefore the corresponding operators are even endomorphisms of $\mathbb{C}^{p|q}$

$$\{\mathbf{T}^{(\bar{0})}\} \in \mathrm{End}(\mathbb{C}^{p|q}), \tag{3.18}$$

but

$$\mathrm{M}^{(\bar{1})}\mathrm{M}'^{(\bar{1})} = \mathrm{M}''^{(\bar{0})}. \tag{3.19}$$

Innovation 3.4. *The set* $\{\mathrm{M}^{(\bar{1})}\}$ *is not closed under composition (matrix multiplication), therefore the corresponding odd operators* $\mathbf{T}^{(1)}$ *are not morphisms by definition at all*

$$\{\mathbf{T}^{(\bar{1})}\} \notin \mathrm{End}(\mathbb{C}^{p|q}). \tag{3.20}$$

Both "even" and "odd" superoperators (considered together) are morphisms.

These considerations should be taken into account during consistent calculations with superqubits and supersymmetric quantum gates.

We remind some common notions in present notations for self-consistency. First, in contrast to the standard transpose operator T: $\mathrm{Mat}(p, \mathbb{C}) \to \mathrm{Mat}(p, \mathbb{C})$, the supertranspose operator sT: $\mathrm{Mat}(p|q, \Lambda_N(\mathbb{C})) \to \mathrm{Mat}(p|q, \Lambda_N(\mathbb{C}))$ is double-valued depending of the parity of supermatrix

$$\begin{pmatrix} A_{p\times p} & B_{p\times q} \\ C_{q\times p} & D_{q\times q} \end{pmatrix}^{\mathsf{sT}} = \begin{cases} \begin{pmatrix} A_{p\times p}^{\mathsf{T}} & C_{q\times p}^{\mathsf{T}} \\ -B_{p\times q}^{\mathsf{T}} & D_{q\times q}^{\mathsf{T}} \end{pmatrix}, & \text{if } \deg \mathrm{M} = \bar{0}, \\ \begin{pmatrix} A_{p\times p}^{\mathsf{T}} & -C_{q\times p}^{\mathsf{T}} \\ B_{p\times q}^{\mathsf{T}} & D_{q\times q}^{\mathsf{T}} \end{pmatrix}, & \text{if } \deg \mathrm{M} = \bar{1}. \end{cases} \tag{3.21}$$

It is seen that $(\mathsf{sT})^{\circ 2} \neq \mathrm{id}$ (while $\mathsf{T}^{\circ 2} = \mathrm{id}$) but $(\mathsf{sT})^{\circ 4} = \mathrm{id}$, and therefore the supertranspose operator is the reflection of order 4, while the transpose is the ordinary reflection (of order 2). For two supermatrices of the same shape $\mathrm{M}, \mathrm{N} \in \mathrm{Mat}(p|q, \Lambda_N(\mathbb{C}))$, we have

$$(\mathrm{MN})^{\mathsf{sT}} = (-1)^{\deg \mathrm{M} \deg N} \mathrm{N}^{\mathsf{sT}} \mathrm{M}^{\mathsf{sT}}, \tag{3.22}$$

and in particular

$$(a\mathrm{M})^{\mathsf{sT}} = a\mathrm{M}^{\mathsf{sT}}, \quad \forall a \in \Lambda_N(\mathbb{C}), \tag{3.23}$$

which means that supertranspose sT is a $\Lambda_N(\mathbb{C})$-module, e.g., in case of the ordinary transpose operator T, which is a $\mathbb{C}$-module.

The supertrace is the homomorphism str: $\mathrm{Mat}(p|q, \Lambda_N(\mathbb{C})) \to \Lambda_N(\mathbb{C})$ that is also double-valued (depending of parity of supermatrix) mapping (for the supermatrix of the standard format (3.11))

$$\mathrm{str}\begin{pmatrix} A_{p\times p} & B_{p\times q} \\ C_{q\times p} & D_{q\times q} \end{pmatrix} = \begin{cases} \mathrm{tr}\, A_{p\times p} - \mathrm{tr}\, D_{q\times q}, & \text{if } \deg \mathrm{M} = \bar{0}, \\ \mathrm{tr}\, A_{p\times p} + \mathrm{tr}\, D_{q\times q}, & \text{if } \deg \mathrm{M} = \bar{1}, \end{cases} \tag{3.24}$$

which is additive and has the supercommutativity property, where M, N are ordinary matrices, and str is invariant with respect to supertranspose (analogous to ordinary trace)

$$\mathrm{tr}\, \mathrm{M}^{\mathsf{T}} = \mathrm{tr}\, \mathrm{M} \overset{\text{susy}}{\Longrightarrow} \mathrm{str}\, \mathrm{M}^{\mathsf{sT}} = \mathrm{str}\, \mathrm{M}. \tag{3.25}$$

The standard superdeterminant (Pakhomov 1974, Berezin and Leites 1975) (or Berezinian Berezin 1987, Leites (1980)) in our notation is

$$\mathrm{Ber}\, \mathrm{M} = \mathrm{Ber}\begin{pmatrix} A_{p\times p} & B_{p\times q} \\ C_{q\times p} & D_{q\times q} \end{pmatrix} = \det\left(A_{p\times p} - B_{p\times q} D_{q\times q}^{-1} C_{q\times p}\right)\left(\det D_{q\times q}\right)^{-1}, \tag{3.26}$$

which differs from the ordinary determinant by the power (-1) in the last multiplier. The mapping Ber is a homomorphism of $\mathrm{Mat}(p|q, \Lambda_N(\mathbb{C}))$ and invariant with respect to supertranspose sT (3.21)

$$\det(\mathrm{M}^{\mathsf{T}}) = \det \mathrm{M} \overset{\text{susy}}{\Longrightarrow} \mathrm{Ber}(\mathrm{M}^{\mathsf{sT}}) = \mathrm{Ber}\, \mathrm{M}. \tag{3.27}$$

The connection of Ber and str is similar to the ordinary case

$$\det \mathrm{M} = e^{\mathrm{tr}(\ln \mathrm{M})} \overset{\text{susy}}{\Longrightarrow} \mathrm{Ber}\, \mathrm{M} = e^{\mathrm{str}(\ln \mathrm{M})}, \tag{3.28}$$

$$\det e^{\mathrm{M}} = e^{\mathrm{tr}\, \mathrm{M}} \overset{\text{susy}}{\Longrightarrow} \mathrm{Ber}\, e^{\mathrm{M}} = e^{\mathrm{str}\, \mathrm{M}}, \tag{3.29}$$

where $\mathrm{M} \in \mathrm{Mat}(p, \mathbb{C})$ and $\mathrm{M} \in \mathrm{Mat}(p|q, \Lambda_N(\mathbb{C}))$.

The Berezinian (3.26) has the inconvenient property for characterizing the entanglement: Ber is not defined for noninvertible $D_{q\times q}$. Therefore, in Borsten *et al* (2010) it was proposed to use another possible function for the entanglement measure, which has many properties of Berezinian (but not all of them), and which satisfies (3.27) and has the ordinary det as the nonsupersymmetric limit (when odd variables vanish). Because the notion sdet is widely used for Ber (Berezin 1987, Leites 1980), we denote this function sdTr, which can be defined by the following informal analogy

$$\det \mathrm{M} = \frac{1}{2}\mathrm{tr}((\mathrm{M}E_{\mathrm{sl}})^{\mathrm{T}}(\mathrm{M}E_{\mathrm{sl}})) \overset{\text{susy}}{\Longrightarrow} \mathrm{sdTr}\, \mathrm{M} = \frac{1}{2}\mathrm{str}\Big(\big(\mathrm{M}E_{\mathrm{osp}}\big)^{\mathrm{sT}}\big(\mathrm{M}E_{\mathrm{osp}}\big)\Big), \tag{3.30}$$

where E_{sl} and $\mathrm{E}_{\mathrm{osp}}$ are $SL(2)$ and $OSp(1|2)$ invariant tensors (Bernstein *et al* 2013) (determining the corresponding group and supergroup in the standard way $M^{\mathrm{T}}E_{\mathrm{sl}}M = M$ and $\mathrm{M}^{\mathrm{sT}}\mathrm{E}_{\mathrm{osp}}\mathrm{M} = \mathrm{M}$). The main property of sdTr is vanishing on the direct product states, and therefore it can measure whether a quantum (two superqubit) state is unentangled or entangled (see below).

3.2 Super Hilbert spaces and operators

Let us denote vectors (quantum states) in the r-dimensional complex Hilbert space $\mathcal{H}_r$ by kets $|\,\psi\rangle$ and the inner product by $\langle|\rangle$: $\mathcal{H}_r \times \mathcal{H}_r \to \mathbb{C}$, which is a nondegenerate Hermitean and positive form that is linear in the first argument and antilinear (conjugate linear) in the second argument. The bra $\langle\varphi\,|$ is defined as an element of the dual space $\mathcal{H}_r^{\dagger}$ (in the notation of Borsten *et al* (2010), for a inner product vector space $\mathcal{V}$ the notation $\mathcal{V}*$ is also used for its dual), which is the functional $\langle\varphi|.\rangle$: $\mathcal{H}_r \to \mathbb{C}$, such that the action on a ket is denoted by $\langle\varphi\,|\,(|\,\psi\rangle) := \langle\varphi|\psi\rangle$ and coincides with the inner product after the identification of the Hilbert space with its dual (Riesz representation theorem Rudin 1991). Informally, one can write the injection $(|\,\psi\rangle)^{\dagger} = \langle\psi\,|$, which in the matrix representation (and finite-dimensional) standardly coincides with Hermitean adjoint, when $|\,\psi\rangle$ becomes a matrix-column, $\langle\psi\,|$ is a matrix-row, and the inner product is a scalar product, the obvious property $\langle\varphi|\psi\rangle^{\dagger} = \langle\psi|\varphi\rangle$ also holds valid.

In a similar way, we consider the $(r|s)$-dimensional super Hilbert space $\mathscr{H}_{(r|s)}$ over $\Lambda_N(\mathbb{C})$ as the $\mathbb{Z}_2$-graded space $\mathscr{H}_{(r|s)} = \mathscr{H}^{(\bar{0})}_{(r|s)} \oplus \mathscr{H}^{(\bar{1})}_{(r|s)}$, such that the supersymmetric quantum states, if homogenous, carry $\mathbb{Z}_2$-grading $\pi_\psi = \bar{0}, \bar{1} \in \mathbb{Z}_2$, and they are denoted by super-kets $||\boldsymbol{\psi}^{(\pi_\psi)}\rangle \in \mathcal{H}^{(\pi_\psi)}_{(r|s)}$ with $\pi_\psi = \deg \boldsymbol{\psi}$, while the body of even super states are the ordinary kets

$$||\boldsymbol{\psi}^{(\bar{0})}\rangle_{\text{body}} = |\,\psi\rangle \in \mathcal{H}_r. \tag{3.31}$$

We denote the super inner product by $\langle||\rangle\colon \mathscr{H}_{(r|s)} \times \mathscr{H}_{(r|s)} \to \Lambda_N(\mathbb{C})$ obeying the property

$$\langle.||.\rangle_{\text{body}} = \langle.|.\rangle \in \mathbb{C}. \tag{3.32}$$

The super dual Hilbert space $\mathscr{H}^{\ddagger}_{(r|s)}$ is defined as the space of the functionals $\langle\boldsymbol{\phi}^{(\pi_\varphi)}||\rangle\colon \mathscr{H}^{(\pi_\psi)}_{(r|s)} \to \Lambda_N(\mathbb{C})$, and the super bra $\langle\boldsymbol{\phi}^{(\pi_\varphi)}||$ with $\pi_\varphi = \deg \boldsymbol{\phi} \in \mathbb{Z}_2$ is given by the action

$$\langle\varphi \mid (|\,\psi\rangle) = \langle\varphi|\psi\rangle \in \mathbb{C} \overset{\text{susy}}{\Longrightarrow} \langle\boldsymbol{\phi}^{(\pi_\varphi)}||(||\boldsymbol{\psi}^{(\pi_\psi)}\rangle) = \delta_{\pi_\varphi\pi_\psi}\langle\boldsymbol{\phi}^{(\pi_\varphi)}||\boldsymbol{\psi}^{(\pi_\psi)}\rangle \in \Lambda_N(\mathbb{C}). \tag{3.33}$$

The presence of the delta function in (3.33) means that the commonly used agreement that the graded super vectors of opposite parity are mutually orthogonal

$$\langle\boldsymbol{\phi}^{(\pi_\varphi)}||\boldsymbol{\psi}^{(\pi_\psi)}\rangle = 0, \text{ if } \pi_\varphi \neq \pi_\psi. \tag{3.34}$$

Therefore, in similar expressions we will put

$$\pi_\varphi = \pi_\psi = \pi = \bar{0},\ \bar{1} \in \mathbb{Z}_2. \tag{3.35}$$

In this case, we have

$$\langle\boldsymbol{\phi}^{(\pi)}||\boldsymbol{\psi}^{(\pi)}\rangle^{\sharp} = \langle\boldsymbol{\psi}^{(\pi)}||\boldsymbol{\phi}^{(\pi)}\rangle. \tag{3.36}$$

Thus, informally, one can write

$$(||\boldsymbol{\psi}^{(\pi)}\rangle)^{\ddagger} = \langle\boldsymbol{\psi}^{(\pi)}||, \tag{3.37}$$

which means that ($\ddagger$) does not change parity π, and it is the reflection of order 4, because

$$(||\boldsymbol{\psi}^{(\pi)}\rangle)^{\ddagger\ddagger} = (-1)^{\pi}||\boldsymbol{\psi}^{(\pi)}\rangle, \tag{3.38}$$

$$(||\boldsymbol{\psi}^{(\pi)}\rangle)^{\ddagger\ddagger\ddagger\ddagger} = ||\boldsymbol{\psi}^{(\pi)}\rangle. \tag{3.39}$$

If $z\in\Lambda_N(\mathbb{C})$ has a fixed parity, then its product with super ket and super bra behaves with respect to ($\ddagger$) differently

$$(||\boldsymbol{\psi}^{(\pi)}\rangle z)^{\ddagger} = (-1)^{\pi \deg z} z^{\sharp}\langle\boldsymbol{\psi}^{(\pi)}||, \tag{3.40}$$

$$(z\langle\boldsymbol{\psi}^{(\pi)}||)^{\ddagger} = (-1)^{\pi(\deg z+1)}||\boldsymbol{\psi}^{(\pi)}\rangle z^{\sharp}, \tag{3.41}$$

where ($\sharp$) is the graded involution or superstar (3.4).

Innovation 3.5. *We can omit the δ-function in (3.33), and this will lead to a new kind of Hilbert spaces that allow mixing of gradings, such that all above formulas should be changed.*

In the super Hilbert space $\mathscr{H}_{(r|s)}$ the superadjoint (‡) of the superoperator (3.9) with the standard graded structure (3.10) is defined by

$$\langle T\ \varphi|\psi\rangle = \langle\varphi|T^{\dagger}\ \psi\rangle \overset{\text{susy}}{\Longrightarrow} \langle \mathbf{T}^{(\pi_T)}\ \boldsymbol{\phi}^{(\pi_\varphi)}||\boldsymbol{\psi}^{(\pi_\psi)}\rangle = (-1)^{\pi_\varphi \pi_T}\langle\boldsymbol{\phi}^{(\pi_\varphi)}||\mathbf{T}^{(\pi_T)\ddagger}\ \boldsymbol{\psi}^{(\pi_\psi)}\rangle, \quad (3.42)$$

where T and $T^{\dagger}$ are an operator and its adjoint in the Hilbert space $\mathcal{H}_r$, $|\,\psi\rangle, |\,\varphi\rangle \in \mathcal{H}_r$, and $\boldsymbol{\phi}^{(\pi_\varphi)} \in \mathcal{H}_{(r|s)}^{(\pi)}$, $\boldsymbol{\psi}^{(\pi_\psi)} \in \mathcal{H}_{(r|s)}^{(\pi_\psi)}$, $\pi_T = \deg \mathbf{T} \in \mathbb{Z}_2$. Here we do not have the restriction (3.35) because the superoperator $\mathbf{T}^{(\pi_T)}$ with $n_T = \bar{1}$ can change the parity of quantum states. The superadjoint of the action on the quantum state is

$$(T\ |\,\psi\rangle)^{\dagger} = \langle\psi|\ \ T^{\dagger} \overset{\text{susy}}{\Longrightarrow} (\mathbf{T}^{(\pi_T)}||\boldsymbol{\psi}^{(\pi_\psi)}\rangle)^{\ddagger} = (-1)^{\pi_\psi \pi_T}\langle\boldsymbol{\psi}^{(\pi_\psi)}||\mathbf{T}^{(\pi_T)\ddagger}, \quad (3.43)$$

The definition (3.42) is equivalent to (see (4.12))

$$\langle\boldsymbol{\phi}^{(\pi_\varphi)}||\mathbf{T}^{(\pi_T)\ddagger}||\boldsymbol{\psi}^{(\pi_\psi)}\rangle = (-1)^{\pi_\varphi\pi_\psi+\pi_\psi+(\pi_\varphi+\pi_\psi)\pi_T}\langle\boldsymbol{\psi}^{(\pi_\psi)}||\mathbf{T}^{(\pi_T)}||\boldsymbol{\phi}^{(\pi_\varphi)}\rangle^{\sharp}. \quad (3.44)$$

If the superoperator $\mathbf{T}$ has a supermatrix representation in $\mathrm{Mat}(p|q,\ \Lambda_N(\mathbb{C}))$, then its superadjoint is represented by composition of supertranspose (3.21) and the graded involution (superstar) (3.4) as

$$\mathrm{M}^{\dagger} = \bar{\mathrm{M}}^{\mathrm{T}},\ \ \mathrm{M} \in \mathrm{Mat}(p,\ \mathbb{C}) \overset{\text{susy}}{\Longrightarrow} \mathrm{M}^{\ddagger} = (\mathrm{M}^{\sharp})^{\mathrm{sT}},\ \mathrm{M} \in \mathrm{Mat}(p|q,\ \Lambda_N(\mathbb{C})), \quad (3.45)$$

which is the superanalog of the Hermitean conjugation (conjugate transpose).

3.3 Qubits and superqubits

Mathematically, qubits (or d-qudits) and superqubits (or $(r|s)$-superqudits) are normalized vectors in the r-dimensional Hilbert space and $(r|s)$-dimensional super Hilbert space, respectively, which are presented in the Dirac bra-ket notation (see previous section). They are written in the computational basis to thoroughly study various symmetries and introduce suitable variables that can consistently measure entanglement. Because the super Hilbert space is $\mathbb{Z}_2$-graded, there can exist even and odd vectors (as for the general quantum states in the previous section) that can correspond to even and odd superqubits, respectively.

The definitions of a single qudit in the complex Hilbert space $\mathcal{H}_d$ and a single superqudit in super Hilbert space $\mathscr{H}_{(r|s)}$ (over $\Lambda_N(\mathbb{C})$) can be written, in general, as the expansions of the (pure) quantum states on the computational (super) basis as follows

$$|\,\Psi\rangle = |\,\Psi\rangle_{(d)} = x_0\,|\,0\rangle + x_1\,|\,1\rangle + \cdots + x_{d-1}\,|\,d-1\rangle, \quad (3.46)$$

$$|\,x\,|_0^2 + |\,x\,|_1^2 + \cdots + |\,x\,|_{d-1}^2 = 1, \quad x_i \in \mathbb{C},\ |\,i\rangle \in \mathcal{H}_d,\ i = 0,\ \ldots,\ d-1, \quad (3.47)$$
$$\Downarrow_{\text{susy}}$$

$$||\mathbf{\Psi}^{(\bar{0})}\rangle = ||\mathbf{0}\rangle x_0 + ||\mathbf{1}\rangle x_1 + \cdots + ||\mathbf{r} - 1\rangle x_{r-1} + ||œ_0\rangle æ_0 + \cdots + ||œ_{s-1}\rangle æ_{s-1}, \quad (3.48)$$

$$x_0^{\sharp} x_0 + x_1^{\sharp} x_1 + \cdots + x_{r-1}^{\sharp} x_{r-1} - æ_0^{\sharp} æ_0 - \cdots æ_{s-1}^{\sharp} æ_{s-1} = 1, \quad (3.49)$$

$$\begin{gathered} ||\mathbf{\Psi}^{(\bar{1})}\rangle = ||\mathbf{0}\rangle æ_0 + ||\mathbf{1}\rangle æ_1 + \cdots + ||\mathbf{r} - 1\rangle æ_{r-1} + ||œ_0\rangle x_0 + \cdots + ||œ_{s-1}\rangle x_{s-1}, \\ x_i \in \Lambda_N^{(0)}(\mathbb{C}),\ ||\mathbf{j}\rangle \in \mathscr{H}_{(r|s)}^{(\bar{0})},\ æ_\alpha \in \Lambda_N^{(\bar{1})}(\mathbb{C}),\ ||œ_\alpha\rangle \in \mathscr{H}_{(r|s)}^{(\bar{1})}. \end{gathered} \quad (3.50)$$

We assume that $\deg ||\mathbf{i}\rangle = \deg \mathbf{x}_i = \bar{0}$, $\deg ||œ_\alpha\rangle = \deg æ_\alpha = \bar{1}$, and therefore the superqudit (3.48) has the even parity $\pi_\Psi = \deg ||\mathbf{\Psi}\rangle = \bar{0}$, and we call it the even superqudit $||\mathbf{\Psi}^{(\bar{0})}\rangle = ||\mathbf{\Psi}\rangle^{\text{even}}$, while the superqudit (3.50) has the odd parity $\pi_\Psi = \deg ||\mathbf{\Psi}\rangle = \bar{1}$, and we call it the odd superqudit $||\mathbf{\Psi}^{(\bar{1})}\rangle = ||\mathbf{\Psi}\rangle^{\text{odd}}$, denoting both of them $||\mathbf{\Psi}^{(\pi)}\rangle = ||\mathbf{\Psi}^{(\pi_\Psi)}\rangle$. The normalization of the odd superqudit can be done using some special Grassmann norms considered in Rudolph (2000), Rogers (2007) and Haba and Kupsch (1995).

Definition 3.6. The qudits $|\,\Psi\rangle$ and superqudits $||\mathbf{\Psi}\rangle$ are

(1) Linear spans of the corresponding subspace $\operatorname{span}(\{|\,i\rangle\}) \subseteq \mathcal{H}_d$ and sub-superspace $\operatorname{span}(\{||\mathbf{i}\rangle\}|\{||œ_\alpha\rangle\}) \subseteq \mathscr{H}_{(r|s)}$, respectively,
(2) Having the normalization conditions (3.47), (3.49).

For consistency, it is natural to assume that the superqudit (3.48) has the Grassmannless limit, body map (Rogers 1980), as the ordinary qudit (3.46)

$$||\mathbf{\Psi}^{(\bar{0})}\rangle_{\text{body}} = |\,\Psi\rangle_{(r)}. \quad (3.51)$$

The normalization conditions (3.47) and (3.49) distinguish (super)qudits among general span subspaces, which allows us to endow them probablistic interpretation. If the limit (3.51) is accepted, then (3.47) and (3.49), as well as the bases $\{|\,i\rangle\} \in \mathcal{H}_r$ and $\{||\mathbf{i}\rangle\} \in \mathscr{H}_{(r|s)}$ are connected with the body map.

The (super)qudits in minimum dimensions $d = 2$, $r = 2$, $s = 1$ are called (super) qubits (Borsten *et al* 2010) and have the form[1]

$$\begin{gathered} |\,\Psi\rangle = |\,\Psi\rangle_{(2)} = x_0\,|\,0\rangle + x_1\,|\,1\rangle, \\ |\,x_0\,|^2 + |\,x_1\,|^2 = 1, \\ x_0, x_1 \in \mathbb{C}, \quad |\,0\rangle, |\,1\rangle \in \mathcal{H}_2, \end{gathered} \quad \overset{\text{susy}}{\Longrightarrow} \quad \begin{gathered} ||\mathbf{\Psi}^{(\bar{0})}\rangle = ||\mathbf{0}\rangle x_0 + ||\mathbf{1}\rangle x_1 + ||œ\rangle æ, \\ x_0^{\sharp} x_0 + x_1^{\sharp} x_1 - æ^{\sharp} æ = 1, \\ ||\mathbf{\Psi}^{(\bar{1})}\rangle = ||\mathbf{0}\rangle æ_0 + ||\mathbf{1}\rangle æ_1 + ||œ\rangle x, \\ x, x_0, x_1 \in \Lambda_N^{(\bar{0})}(\mathbb{C}), \quad ||\mathbf{0}\rangle, ||\mathbf{1}\rangle \in \mathscr{H}_{(2|1)}^{(\bar{0})}, \\ æ, æ_0, æ_1 \in \Lambda_N^{(\bar{1})}(\mathbb{C}), \quad ||œ\rangle \in \mathscr{H}_{(2|1)}^{(\bar{1})}. \end{gathered} \quad (3.52)$$

[1] For clarity and convenience for applications, we use the manifest presentation of different variables. The right coordinates are used in superqubits according to the sign agreement of Borsten *et al* (2010).

There are four main operations between two single (super)qubits.

(1) *Inner product* of bra (super)qubit and ket (super)qubit

$$\langle\Psi|\Psi'\rangle = \bar{x}_0 x_0' + \bar{x}_1 x_1' \in \mathbb{C} \overset{\text{susy}}{\Longrightarrow} \begin{array}{l} \langle\mathbf{\Psi}^{(\bar{0})}||\mathbf{\Psi}^{(\bar{0})\prime}\rangle = x_0^{\sharp}x_0' + x_1^{\sharp}x_1' - æ^{\sharp}æ', \\ \langle\mathbf{\Psi}^{(\bar{1})}||\mathbf{\Psi}^{(\bar{1})\prime}\rangle = æ_0^{\sharp}æ_0' + æ_1^{\sharp}æ_1' + x^{\sharp}x' \end{array} \tag{3.53}$$

where $\overline{()}$ is complex conjugation and ($\sharp$) is the grade involution (3.4).

If the states coincide, $|\Psi'\rangle = |\Psi\rangle$ and $||\mathbf{\Psi}'\rangle = ||\mathbf{\Psi}\rangle$, then (3.53) are square norms of $|\Psi\rangle$ and $||\mathbf{\Psi}^{(\bar{0})}\rangle$ becoming unity for normalized (super)qubits. For physical states, the square norm of the even superqubit is positive

$$||\mathbf{\Psi}^{(\bar{0})}||^2_{\text{body}} = \langle\mathbf{\Psi}^{(\bar{0})}||\mathbf{\Psi}^{(\bar{0})\prime}\rangle_{\text{body}} > 0. \tag{3.54}$$

(2) *Outer product* of ket and bra gives the density (super)matrix of (super)qubit

$$\rho = |\Psi\rangle\langle\Psi| = \begin{pmatrix} x_0\bar{x}_0 & x_1\bar{x}_0 \\ x_0\bar{x}_1 & x_1\bar{x}_1 \end{pmatrix} \tag{3.55}$$

$$\Downarrow_{\text{susy}} \tag{3.56}$$

$$\rho^{(\bar{0})} = ||\mathbf{\Psi}^{(\bar{0})}\rangle\langle\mathbf{\Psi}^{(\bar{0})}|| = \left(\begin{array}{ccc} x_0x_0^{\sharp} & x_1x_0^{\sharp} & æx_0^{\sharp} \\ x_0x_1^{\sharp} & x_1x_1^{\sharp} & æx_1^{\sharp} \\ \hline -x_0æ^{\sharp} & -x_1æ^{\sharp} & -ææ^{\sharp} \end{array}\right), \tag{3.57}$$

$$\rho^{(\bar{1})} = ||\mathbf{\Psi}^{(\bar{1})}\rangle\langle\mathbf{\Psi}^{(\bar{1})}|| = \left(\begin{array}{ccc} -æ_0æ_0^{\sharp} & -æ_1æ_0^{\sharp} & -xæ_0^{\sharp} \\ -æ_0æ_1^{\sharp} & -æ_1æ_1^{\sharp} & -xæ_1^{\sharp} \\ \hline æ_0x^{\sharp} & æ_1x^{\sharp} & xx^{\sharp} \end{array}\right), \tag{3.58}$$

and the body map limit for $\rho^{(\bar{0})}$ is similar to (3.51). The standard connection of the inner product with the (super)trace of density matrix for a given (super)qubit holds valid (taking into account gradings)

$$\operatorname{tr}\rho = \langle\Psi|\Psi\rangle \in \mathbb{C} \overset{\text{susy}}{\Longrightarrow} \begin{array}{l} \operatorname{str}\rho^{(\bar{0})} = \langle\mathbf{\Psi}^{(\bar{0})}||\mathbf{\Psi}^{(\bar{0})}\rangle \in \Lambda_N^{(\bar{0})}(\mathbb{C}), \\ \operatorname{str}\rho^{(\bar{1})} = -\langle\mathbf{\Psi}^{(\bar{1})}||\mathbf{\Psi}^{(\bar{1})}\rangle \in \Lambda_N^{(\bar{0})}(\mathbb{C}). \end{array} \tag{3.59}$$

(3) *Tensor product* of two ket (super)qubits (or two bra (super)qubits) (3.52) can be presented as the following manifest expansions on elementary tensors (all gradings appear and are shown for clarity and direct usage in computations)

$$
\begin{aligned}
&\mid \Psi\rangle \otimes \mid \Psi'\rangle = x_0 x_0' \mid 0\rangle \otimes \mid 0'\rangle + x_0 x_1' \mid 0\rangle \otimes \mid 1'\rangle + x_1 x_0' \mid 1\rangle \otimes \mid 0'\rangle + x_1 x_1' \mid 1\rangle \otimes \mid 1'\rangle \\
&\Downarrow_{\text{susy}} \\
&\begin{pmatrix} ||\mathbf{\Psi}^{(\bar{0})}\rangle \otimes ||\mathbf{\Psi}^{(\bar{0})\prime}\rangle \\ ||\mathbf{\Psi}^{(\bar{1})}\rangle \otimes ||\mathbf{\Psi}^{(\bar{1})\prime}\rangle \\ ||\mathbf{\Psi}^{(\bar{0})}\rangle \otimes ||\mathbf{\Psi}^{(\bar{1})\prime}\rangle \\ ||\mathbf{\Psi}^{(\bar{1})}\rangle \otimes ||\mathbf{\Psi}^{(\bar{0})\prime}\rangle \end{pmatrix} = ||\mathbf{0}\rangle \otimes ||\mathbf{0}'\rangle \begin{pmatrix} x_0 x_0' \\ \text{æ}_0 \text{æ}_0' \\ x_0 \text{æ}_0' \\ \text{æ}_0 x_0' \end{pmatrix} + ||\mathbf{0}\rangle \otimes ||\mathbf{1}'\rangle \begin{pmatrix} x_0 x_1' \\ \text{æ}_0 \text{æ}_1' \\ x_0 \text{æ}_1' \\ \text{æ}_0 x_1' \end{pmatrix} \\
&+ ||\mathbf{0}\rangle \otimes ||\text{œ}'\rangle \begin{pmatrix} x_0 \text{æ}' \\ \text{æ}_0 x' \\ x_0 x' \\ \text{æ}_0 \text{æ}' \end{pmatrix} + ||\mathbf{1}\rangle \otimes ||\mathbf{0}'\rangle \begin{pmatrix} x_1 x_0' \\ \text{æ}_1 \text{æ}_0' \\ x_1 \text{æ}_0' \\ \text{æ}_1 x_0' \end{pmatrix} + ||\mathbf{1}\rangle \otimes ||\mathbf{1}'\rangle \begin{pmatrix} x_1 x_1' \\ \text{æ}_1 \text{æ}_1' \\ x_1 \text{æ}_1' \\ \text{æ}_1 x_1' \end{pmatrix} \\
&+ ||\mathbf{1}\rangle \otimes ||\text{œ}'\rangle \begin{pmatrix} x_1 \text{æ}' \\ \text{æ}_1 x' \\ x_1 x' \\ \text{æ}_1 \text{æ}' \end{pmatrix} + ||\text{œ}\rangle \otimes ||\mathbf{0}'\rangle \begin{pmatrix} \text{æ} x_0' \\ x \text{æ}_0' \\ \text{æ}\text{æ}_0' \\ x x_0' \end{pmatrix} + ||\text{œ}\rangle \otimes ||\mathbf{1}'\rangle \begin{pmatrix} \text{æ} x_1' \\ x \text{æ}_1' \\ \text{æ}\text{æ}_1' \\ x x_1' \end{pmatrix} \\
&+ ||\text{œ}\rangle \otimes ||\text{œ}'\rangle \begin{pmatrix} \text{æ}\text{æ}' \\ x x' \\ \text{æ} x' \\ x \text{æ}' \end{pmatrix} \in \begin{pmatrix} \left(\mathscr{H}_{(2|1)} \otimes \mathscr{H}_{(2|1)}\right)^{(\bar{0})} \\ \left(\mathscr{H}_{(2|1)} \otimes \mathscr{H}_{(2|1)}\right)^{(\bar{0})} \\ \left(\mathscr{H}_{(2|1)} \otimes \mathscr{H}_{(2|1)}\right)^{(\bar{1})} \\ \left(\mathscr{H}_{(2|1)} \otimes \mathscr{H}_{(2|1)}\right)^{(\bar{1})} \end{pmatrix}.
\end{aligned}
\tag{3.60}
$$

Thus, there are four different superqubit tensor products, depending of their parity.

Definition 3.7. The (pure) quantum state that can be obtained as a tensor product is called a separable state.

(4) *Cross product* of two ket qutrits ((3.46) with $d = 3$) of the form

$$\mid \Psi\rangle = \mid \Psi\rangle_{(3)} = x_0 \mid 0\rangle + x_1 \mid 1\rangle + x_2 \mid 2\rangle, \tag{3.61}$$

$$
\begin{aligned}
&\mid x_0 \mid^2 + \mid x_1 \mid^2 + \mid x_2 \mid^2 = 1, \\
&x_0, x_1, x_2 \in \mathbb{C}, \quad \mid 0\rangle, \mid 1\rangle, \mid 2\rangle \in \mathcal{H}_3,
\end{aligned}
\tag{3.62}
$$

can be defined by analogy with ordinary cross product of vectors

$$\mid \Phi\rangle_{\text{cross}} = \mid \Psi\rangle \times \mid \Psi'\rangle = \sum_{i,j,k=0,1,2} \epsilon_{ijk} x_j x_k' \mid i\rangle \tag{3.63}$$

$$= \det \begin{pmatrix} \mid 0\rangle & \mid 1\rangle & \mid 2\rangle \\ x_0 & x_1 & x_2 \\ x_0' & x_1' & x_2' \end{pmatrix} = \det M_{\mid 0\rangle} \mid 0\rangle + \det M_{\mid 1\rangle} \mid 1\rangle + \det M_{\mid 2\rangle} \mid 2\rangle \tag{3.64}$$

$$= (x_1 x_2' - x_2 x_1') \mid 0\rangle - (x_0 x_2' - x_2 x_0') \mid 1\rangle + (x_0 x_1' - x_1 x_0') \mid 2\rangle, \quad x_i, x_i' \in \mathbb{C}, \quad (3.65)$$

where $M_{|i\rangle}$ is the minor of element $| i\rangle$, and ϵ_{ijk} fully antisymmetric tensor. The last expanded form (3.65) is convenient to use for superqubits as well.

Definition 3.8. We call the qutrit $| \Phi\rangle_{\mathrm{cross}}$ that is built as the cross product (3.65) a cross-qutrit.

The square norm of the cross-qutrit is

$$\begin{aligned} \|| \Phi\rangle_{\mathrm{cross}} \|^2 &= \|\Psi\|^2 \|\Psi'\|^2 - |\langle\Psi|\Psi'\rangle|^2 \\ &= \left| \det \mathrm{M}_{|0\rangle} \right|^2 + \left| \det \mathrm{M}_{|1\rangle} \right|^2 + \left| \det \mathrm{M}_{|2\rangle} \right|^2. \end{aligned} \quad (3.66)$$

Therefore, for the normalized qutrit $| \Phi\rangle_{\mathrm{cross}}$, we have the additional condition (together with two conditions (3.62) for $| \Psi\rangle$ and $| \Psi'\rangle$)

$$\left| \det \mathrm{M}_{|0\rangle} \right|^2 + \left| \det \mathrm{M}_{|1\rangle} \right|^2 + \left| \det \mathrm{M}_{|2\rangle} \right|^2 = 1. \quad (3.67)$$

Definition 3.9. The (pure) quantum state which can be obtained as a cross product is called a cross-separable state.

The cross-qutrits have special properties and can be connected with the concurrence measure in considering entanglement (see below).

3.4 Multi-(super)qubit states

The multi-(super)qudit quantum states are vectors in the tensor product of n (super) Hilbert spaces $\mathcal{H}_d^{\otimes n} = \overbrace{\mathcal{H}_d \otimes \cdots \otimes \mathcal{H}_d}^{n}$ (resp. $\mathscr{H}_{(r|s)}^{\otimes n} = \overbrace{\mathscr{H}_{(r|s)} \otimes \cdots \otimes \mathscr{H}_{(r|s)}}^{n}$). On first sight, a straightforward way to obtain such vectors is to use the tensor product (3.60) repeatedly $n - 1$ times. However, this procedure is too restricted and can only give separable states. The consequent definition should be made in terms of spans, as in **Definition** 3.6.

Definition 3.10. The multi-qudits (n-qudit states) are

(1) Linear span of the Hilbert subspace

$$\begin{aligned} &\{| \Psi(n)\rangle\} = \mathrm{span}(| i_1\rangle \otimes \cdots \otimes | i_n\rangle) \subseteq \overbrace{\mathcal{H}_d \otimes \cdots \otimes \mathcal{H}_d}^{n}, \quad | i_k\rangle \in \mathcal{H}_d, \quad k = 1, \ldots, n, \\ &| \Psi(n)\rangle = \sum_{i_1=0}^{d-1} \cdots \sum_{i_n=0}^{d-1} x_{i_1 \cdots i_n} | i_1\rangle \otimes \cdots \otimes | i_n\rangle, \quad x_{i_1 \cdots i_n} \in \mathbb{C}, \end{aligned} \quad (3.68)$$

(2) With the normalization (3.46)

$$\sum_{i_1=0}^{d-1} \cdots \sum_{i_n=0}^{d-1} | x |^2_{i_1 \cdots i_n} = 1. \tag{3.69}$$

Definition 3.11. The multi-superqudits (n-superqudit states) are

(1) Linear span of the super Hilbert subspace

$$\begin{gathered}\operatorname{span}(||\mathbf{I}_1\rangle \otimes \cdots \otimes ||\mathbf{I}_n\rangle) \subseteq \overbrace{\mathscr{H}_{(r|s)} \otimes \cdots \otimes \mathscr{H}_{(r|s)}}^{n}, \quad ||\mathbf{I}_k\rangle = (||\mathbf{i}_k\rangle, ||œ_k\rangle), \\ ||\mathbf{i}_k\rangle \in \mathscr{H}^{(\bar{0})}_{(r|s)}, \; ||œ_k\rangle \in \mathscr{H}^{(\bar{1})}_{(r|s)}, \quad \mathrm{k} = 1, \ldots, \mathrm{n},\end{gathered} \tag{3.70}$$

which respect parities of variables, such that we have even and odd superqubits (see (3.6))

$$\begin{gathered}||\boldsymbol{\Psi}^{(\mathsf{k})}(n)\rangle = \sum_{j_1=0}^{n-1} \cdots \sum_{j_n=0}^{n-1} \mathbf{y}_{j_1 \cdots j_n} ||\mathbf{I}_1\rangle \otimes \cdots \otimes ||\mathbf{I}_n\rangle, \quad \mathbf{y}_{j_1 \cdots j_n} = (x_{j_1 \cdots j_n}, æ_{j_1 \cdots j_n}), \\ \mathsf{k} = \deg \mathbf{y}_{j_1 \cdots j_n} \boxplus \deg ||\mathbf{I}_1\rangle \boxplus \cdots \deg ||\mathbf{I}_n\rangle = \bar{0}, \bar{1} \in \mathbb{Z}_2,\end{gathered} \tag{3.71}$$

(2) Normalization can be made for the even multi-superqubit $||\boldsymbol{\Psi}^{(\bar{0})}(n)\rangle$ only, as for the single superqubit (3.52).

To clarify the difference between the separable (3.60) and nonseparable (3.68), (3.71) states, we consider the example of two (super)qubits. Thus, for $n = 2$ (two-party states), we obtain

$$\begin{gathered} |\Psi(2)\rangle = x_{00}|0\rangle \otimes |0'\rangle + x_{01}|0\rangle \otimes |1'\rangle + x_{10}|1\rangle \otimes |0'\rangle + x_{11}|1\rangle \otimes |1'\rangle \\ \Downarrow_{\text{susy}} \\ \begin{pmatrix} ||\boldsymbol{\Psi}^{(\bar{0})}(2)\rangle \\ ||\boldsymbol{\Psi}^{(\bar{1})}(2)\rangle \end{pmatrix} = ||\mathbf{0}\rangle \otimes ||\mathbf{0}'\rangle \begin{pmatrix} x_{00} \\ æ_{00} \end{pmatrix} + ||\mathbf{0}\rangle \otimes ||\mathbf{1}'\rangle \begin{pmatrix} x_{01} \\ æ_{01} \end{pmatrix} \\ + ||\mathbf{0}\rangle \otimes ||œ'\rangle \begin{pmatrix} æ_{02} \\ x_{02} \end{pmatrix} + ||\mathbf{1}\rangle \otimes ||\mathbf{0}'\rangle \begin{pmatrix} x_{10} \\ æ_{10} \end{pmatrix} + ||\mathbf{1}\rangle \otimes ||\mathbf{1}'\rangle \begin{pmatrix} x_{11} \\ æ_{11} \end{pmatrix} \\ + ||\mathbf{1}\rangle \otimes ||œ'\rangle \begin{pmatrix} æ_{12} \\ x_{12} \end{pmatrix} + ||œ\rangle \otimes ||\mathbf{0}'\rangle \begin{pmatrix} æ_{20} \\ x_{20} \end{pmatrix} + ||œ\rangle \otimes ||\mathbf{1}'\rangle \begin{pmatrix} æ_{21} \\ x_{21} \end{pmatrix} \\ + ||œ\rangle \otimes ||œ'\rangle \begin{pmatrix} x_{22} \\ æ_{22} \end{pmatrix} \in \begin{pmatrix} \left(\mathscr{H}_{(2|1)} \otimes \mathscr{H}'_{(2|1)}\right)^{(\bar{0})} \\ \left(\mathscr{H}_{(2|1)} \otimes \mathscr{H}'_{(2|1)}\right)^{(\bar{1})} \end{pmatrix}, \end{gathered} \tag{3.72}$$

where $|\Psi(2)\rangle$ has four bosons, $||\boldsymbol{\Psi}^{(\bar{0})}(2)\rangle$ has five bosons and four fermions, $||\boldsymbol{\Psi}^{(\bar{1})}(2)\rangle$ has four bosons and five fermions. Comparing the tensor product (3.60) and (3.72), we observe that for separable states all the amplitudes (coordinates) in (3.72) can be composed

$$\begin{array}{l} x_{ij} \overset{\text{sep}}{=} x_i x'_j, \\ x_i \in \mathbb{C}, \\ i, j = 0, 1 \end{array} \overset{\text{susy}}{\Longrightarrow} \begin{array}{c} x_{ij} \overset{\text{sep}}{=} \begin{cases} x_i x'_j \\ \text{æ}_i \text{æ}'_j \end{cases}, x_{i2} \overset{\text{sep}}{=} \begin{cases} x_i x' \\ \text{æ}_i \text{æ}' \end{cases}, x_{2i} \overset{\text{sep}}{=} \begin{cases} \text{ææ}'_i \\ x x'_i \end{cases}, x_{22} \overset{\text{sep}}{=} \begin{cases} \text{ææ}' \\ x x' \end{cases} \\ \text{æ}_{ij} \overset{\text{sep}}{=} \begin{cases} x_i \text{æ}'_j \\ \text{æ}_i x'_j \end{cases}, \text{æ}_{i2} \overset{\text{sep}}{=} \begin{cases} x_i \text{æ}' \\ \text{æ}_i x' \end{cases}, \text{æ}_{2i} \overset{\text{sep}}{=} \begin{cases} \text{æ} x'_i \\ x \text{æ}'_i \end{cases}, \text{æ}_{22} \overset{\text{sep}}{=} \begin{cases} \text{æ} x' \\ x \text{æ}' \end{cases} \\ x, x_i, x_{ij}, x_{i2}, x_{2i}, x', x'_i \in \Lambda_N^{(\bar{0})}(\mathbb{C}), \\ \text{æ}, \text{æ}_i, \text{æ}_{ij}, \text{æ}_{i2}, \text{æ}_{2i}, \text{æ}', \text{æ}'_i \in \Lambda_N^{(\bar{1})}(\mathbb{C}). \end{array} \tag{3.73}$$

Remark 3.12. The separability of two-party superqubit even $||\boldsymbol{\Psi}^{(\bar{0})}(2)\rangle$ and odd $||\boldsymbol{\Psi}^{(\bar{1})}(2)\rangle$ states (3.72) is determined in the nonunique way (3.73).

Definition 3.13. Multi-(super)qubit states are called entangled (inseparable) if at least one of their amplitudes ($\mathbf{y}_{j_1 \cdots j_n}$ in (3.71)) cannot be presented in the composite factorized form (3.73).

A suitable function that can measure entanglement should have the main property: vanishing for the separable states (3.60). The simplest such function for two qubits (without other requirements) is the determinant. Indeed, for the separable two party (super)qubit system, we have from the factorization (3.73)

$$\begin{array}{l} |\Psi(2)\rangle : f(x) = \det x_{ij} \overset{\text{sep}}{=} \det \left(x_i x'_j \right) \equiv 0, \quad \forall x_i, x'_j \in \mathbb{C}, \ i, j = 0, 1. \\ \Downarrow_{\text{susy}} \end{array} \tag{3.74}$$

$$||\boldsymbol{\Psi}^{(\bar{0})}(2)\rangle : f^{(\bar{0})}(y) = \det (x_{ij} x_{22} + \text{æ}_{i2} \text{æ}_{2j}) \overset{\text{sep}}{=} 0, \ y_{ij} = (x_{ij}, \text{æ}_{ij}), \tag{3.75}$$

$$||\boldsymbol{\Psi}^{(\bar{1})}(2)\rangle : f^{(\bar{1})}(y) = \det (\text{æ}_{ij} \text{æ}_{22} - x_{i2} x_{2j}) \overset{\text{sep}}{=} 0, \ x_{ij} \in \Lambda_N^{(\bar{0})}(\mathbb{C}), \ \text{æ}_{ij} \in \Lambda_N^{(\bar{1})}(\mathbb{C}). \tag{3.76}$$

Further requirements can be imposed, for ordinary qubits they are positivity, monotonicity, and the range in $\{0, 1\}$, as probability, which gives the concurrence (Hill and Wootters 1997, Wootters 1998, Horodecki *et al* 2009)

$$C_2(x) = C(|\Psi(2)\rangle) = 2\,|f(x)| = 2\left|\det x_{ij}\right|, \tag{3.77}$$

such that for maximally entangled states, e.g., the Bell state $x_{11} = x_{22} = \frac{1}{\sqrt{2}}$, $x_{01} = x_{10} = 0$, to get $C(x) = 1$.

Innovation 3.14. *We define the even and odd superconcurrences by*

$$C^{(\bar{0})}(y) = C(||\boldsymbol{\Psi}^{(\bar{0})}(2)\rangle) = 2 \left\| \det\left(x_{ij}\varkappa_{22} + æ_{i2}æ_{2j}\right) \right\|_R, \tag{3.78}$$

$$C^{(\bar{1})}(y) = C(||\boldsymbol{\Psi}^{(\bar{1})}(2)\rangle) = 2 \left\| \det\left(æ_{ij}æ_{22} - x_{i2}\varkappa_{2j}\right) \right\|_R, \tag{3.79}$$

where $\|\|_R$ *is one of the Grassmann norms* (De Witt 1992, Rogers 2007, Rudolph 2000).

The square of the concurrence is called tangle (Borsten *et al* 2010), which can be written for two qubits in the form

$$\tau(x) = 4f(x)\overline{f(x)} = 4 \det x_{ij} \det \bar{x}_{ij}, \tag{3.80}$$

where $(\overline{\_})$ is the complex conjugation.

Innovation 3.15. *For two even/odd superqubits, by analogy with (3.80) and taking into account possible noninvertibilities, we can define the even supertangle* $\boldsymbol{\tau}^{(\bar{0})}(x)$ *and odd supertangle* $\boldsymbol{\tau}^{(\bar{1})}(x)$ *in the following way*

$$\boldsymbol{\tau}^{(\bar{0})}(y)\varkappa_{22}(\varkappa_{22})^{\sharp} = 4f^{(\bar{0})}(y)\left(f^{(\bar{0})}(y)\right)^{\sharp}, \tag{3.81}$$

$$\boldsymbol{\tau}^{(\bar{1})}(y)æ_{22}(æ_{22})^{\sharp} = 4f^{(\bar{1})}(y)\left(f^{(\bar{1})}(y)\right)^{\sharp}, \tag{3.82}$$

where $f^{(\bar{0})}(y)$ *and* $f^{(\bar{1})}(y)$ *are defined in (3.75) and (3.76), respectively.*

In case of invertible $\varkappa_{22}$, the even superconcurrence $C^{(\bar{0})}(y)$ (3.78) and even supertangle $\boldsymbol{\tau}^{(\bar{0})}(x)$ (3.81) can be connected with the Berezinian (3.26).

There are many other entanglement measures, e.g., entropy of entanglement, positive partial transpose, quantum discord, entanglement of formation, distillable entanglement, entanglement cost, squashed entanglement, and entanglement witnesses (Horodecki *et al* 2009). Some of them can also be applied for multi-(super) qudits, for superqubits, see, e.g. Borsten *et al* (2010).

The entanglement classification and manipulation can be provided by considering various local symmetries of multi-(super)qubit systems. The main paradigm is local operations and classical communication (LOCC), which was proposed in Bennett *et al* (1996): it is not possible to change the quantum property of a many party state (e.g. increase its entanglement) using local operations (e.g. on one party qubits) and classical channels only. Thus, the many party quantum states can be classified in such a way that each class contains the representative state with maximum entanglement. If some operations can be performed using LOCC, but may fail, they are called stochastic local operations and classical communication (SLOCC) (Vidal 2000, Verstraete *et al* 2001). Quantum states that can be transformed into one another are called SLOCC equivalent and the corresponding equivalence classes are called entanglement (SLOCC) classes, which are invariant under invertible unitary transformations (Eltschka and Siewert 2014, Verstraete *et al* 2001).

A single qubit $|\Psi\rangle$ (3.52) carries the fundamental representation of $SU(2)$ group, and therefore for the n-qubit state the LOCC equivalence group is $[SU(2,\mathbb{C})]^{\otimes n}$ (Vidal 2000, Verstraete *et al* 2001), while the SLOCC equivalence group is $[SL(2,\mathbb{C})]^{\otimes n}$ (Borsten *et al* 2010). Thus, any separable n-qubit state will remain separable under all $[SU(2)]^{\otimes n}$ operations. In a similar way, the even superqubit $||\boldsymbol{\Psi}^{(\bar{0})}\rangle$ (3.52) carries the fundamental representation of the local operation (unitary orthosymplectic) group $uOSp(2|1)$, and so for n-superqubit state the LOCC equivalence group is $[uOSp(2|1)]^{\otimes n}$ and the SLOCC equivalence group is $[OSp(2|1)]^{\otimes n}$ (Borsten *et al* 2015).

The supersymmetrization of (S)LOCC groups is different from supersymmetrization of the Poincaré group, and therefore artificially adding the superpartners of the electron and photon does not give a superqubit (Brádler 2012). Nevertheless, supersymmetric extension of quantum mechanics based on superqubits may be a candidate for a superquantum theory that lies in the gap between the ordinary quantum theory and nonlocal boxes (Popescu and Rohrlich 1994, Borsten *et al* 2014). There can be applications of superqubits in condensed matter physics where the orthosymplectic Lie superalgebras play an important role (Efetov 1997).

3.5 Innovations

Here we consider the following generalizations of superqubits.

Innovation 3.16. (Odd superqubits). *The odd superqubits $||\boldsymbol{\Psi}^{(\bar{1})}\rangle$ were introduced in (3.52) by analogy with the odd superfields. We suppose that the SLOCC equivalence group for odd superqubits could be connected with the periplectic group, a subgroup of the general linear supergroup over $\Lambda_N(\mathbb{C})$, which preserves the odd bilinear form* (Leites and Serganova 1991, Deligne *et al* 2018).

Innovation 3.17. [Tensor product of qubit and superqubit] *From the first glance, one can think that the tensor product of $|\Psi\rangle \in \mathcal{H}_2$ and $||\boldsymbol{\Psi}'\rangle \in \mathscr{H}_{(2|1)}$ is a particular case of $||\boldsymbol{\Psi}\rangle \otimes ||\boldsymbol{\Psi}'\rangle$ (3.60) where one multiplier is the body map (3.51). However, the consistent construction is more complicated because the spaces $\mathcal{H}_d$ and $\mathscr{H}_{(r|s)}$ are over different fields. In general, the tensor product of the vector space $\mathcal{V}_1$ over $\mathbb{k}_1$ and $\mathcal{V}_2$ over $\mathbb{k}_2$ can be built as*

$$\mathcal{V}_1 \otimes_{\mathbb{k}_2} \mathcal{V}_2 := \mathcal{V}_1 \otimes_{\mathbb{k}_2} \left(\mathbb{k}_2 \otimes_{\mathbb{k}_1} \mathcal{V}_2\right). \tag{3.83}$$

The same construction can be provided for the corresponding Hilbert spaces by consideration their inner products. Moreover, the properties of the qubit tensor product with even $||\mathbf{\Psi}^{(\bar{0})}\rangle$ *and odd* $||\mathbf{\Psi}^{(\bar{1})}\rangle$ *superqubits are fully different. Indeed, from (3.52) we have the mixed qubit-superqubit tensor product (using informally the same its sign)*

$$\begin{pmatrix} |\Psi\rangle \otimes ||\mathbf{\Psi}^{(\bar{0})\prime}\rangle \\ |\Psi\rangle \otimes ||\mathbf{\Psi}^{(\bar{1})\prime}\rangle \end{pmatrix} = |0\rangle \otimes ||\mathbf{0}'\rangle \begin{pmatrix} x_0 x_0' \\ x_0 æ_0' \end{pmatrix} + |0\rangle \otimes ||\mathbf{1}'\rangle \begin{pmatrix} x_0 x_1' \\ x_0 æ_1' \end{pmatrix}$$
$$+ |0\rangle \otimes ||œ'\rangle \begin{pmatrix} x_0 æ' \\ x_0 x' \end{pmatrix} + |1\rangle \otimes ||\mathbf{0}'\rangle \begin{pmatrix} x_1 x_0' \\ x_1 æ_0' \end{pmatrix} + |1\rangle \otimes ||\mathbf{1}'\rangle \begin{pmatrix} x_1 x_1' \\ x_1 æ_1' \end{pmatrix} \tag{3.84}$$
$$+ |1\rangle \otimes ||œ'\rangle \begin{pmatrix} x_1 æ' \\ x_1 x' \end{pmatrix} \in \begin{pmatrix} \left(\mathcal{H}_2 \otimes \mathscr{H}_{(2|1)}\right)^{(\bar{0})} \\ \left(\mathcal{H}_2 \otimes \mathscr{H}_{(2|1)}\right)^{(\bar{1})} \end{pmatrix}.$$

Such mixed tensor products and corresponding qubit-superqubit nonseparable quantum states would be worthwhile to investigate in detail from the viewpoint of entanglement and constructing SLOCC equivalence groups for them.

Innovation 3.18. [Concurrence through cross product of qutrits] *Here we show that concurrence of 2-qutrit states can be expressed through the cross product of qutrits. Let us consider a general nonseparable 2-qutrit state which is not the tensor product of two qutrits (3.61)*

$$|\Psi(2)\rangle_{(3)} = \sum_{i,j'=0,1,2} x_{ij'} \, |i\rangle \otimes |j'\rangle, \tag{3.85}$$

$$\sum_{i,j'=0,1,2} |x|_{ij'}^2 = 1, \quad x_{ij'} \in \mathbb{C}, \quad |i\rangle \in \mathcal{H}_3, |j'\rangle \in \mathcal{H}_3'. \tag{3.86}$$

Now we present 2-qutrit state (3.85) as some special kind of superposition by introducing three (ancilla) qutrits $|\Phi_i\rangle_{(3)}$ *and call it the semi-separable form of 2-qutrit state*

$$|\Psi(2)\rangle_{(3)} = a_0 |\Phi_0\rangle \otimes |0'\rangle + a_1 |\Phi_1\rangle \otimes |1'\rangle + a_2 |\Phi_2\rangle \otimes |2'\rangle, \tag{3.87}$$

$$|\Phi_i\rangle = |\Phi_i\rangle_{(3)} = y_{i0} |0\rangle + y_{i1} |1\rangle + y_{i2} |2\rangle, \quad a_i, y_{ij} \in \mathbb{C}, \ i, j = 0, 1, 2, \tag{3.88}$$

$$\sum_{j=0,1,2} |y|_{ij}^2 = 1, \textit{for each } i = 0, 1, 2, \qquad |i\rangle \in \mathcal{H}_3, |j'\rangle \in \mathcal{H}_3'. \tag{3.89}$$

Both normalizatons (3.86) and (3.89) lead to the restriction on the coefficients a_i *in the expansion (3.87), and if they are real* $a_i \in \mathbb{R}$, *then* $a_0^2 + a_1^2 + a_2^2 = 1$, *moreover, in the simplest case we can choose*

$$a_0 = a_1 = a_2 = \frac{1}{\sqrt{3}}. \tag{3.90}$$

Thus, we obtain the relation between amplitudes

$$y_{ij} = x_{ij}\sqrt{3}. \tag{3.91}$$

Let us construct three cross products (3.65) of the ancilla qutrits (3.88) $\left| \Phi_i \times \Phi_j \right\rangle \equiv | \Phi_i\rangle \times \left| \Phi_j \right\rangle$, $(i, j) = (0, 1), (1, 2), (2, 0)$, *which have the square norms (3.66) (given in terms of their amplitudes (3.88) in the form which is convenient for application to superqubits)*

$$||\Phi_0 \times \Phi_1 ||^2 = \left| y_{00}y_{11} - y_{01}y_{10} \right|^2 + \left| y_{00}y_{12} - y_{02}y_{10} \right|^2 + \left| y_{01}y_{12} - y_{02}y_{11} \right|^2, \tag{3.92}$$

$$||\Phi_1 \times \Phi_2 ||^2 = \left| y_{10}y_{21} - y_{11}y_{20} \right|^2 + \left| y_{10}y_{22} - y_{12}y_{20} \right|^2 + \left| y_{11}y_{22} - y_{12}y_{21} \right|^2, \tag{3.93}$$

$$||\Phi_2 \times \Phi_0 ||^2 = \left| y_{00}y_{21} - y_{01}y_{20} \right|^2 + \left| y_{00}y_{22} - y_{02}y_{20} \right|^2 + \left| y_{01}y_{22} - y_{02}y_{21} \right|^2. \tag{3.94}$$

Observe, that the sum of the square norms after the substitution (3.91) coincides with the concurrence for 2-qutrits (Cereceda 2003) (*in* Pashaev 2023 *the coefficient* $\sqrt{3}$ *was lost). Thus, we obtain the expression for the 2-qutrit concurrence*

$$C_3\left(| \Psi(2)\rangle_{(3)}\right) = \sqrt{||\Phi_0 \times \Phi_1 ||^2 + ||\Phi_1 \times \Phi_2 ||^2 + ||\Phi_2 \times \Phi_0 ||^2} \, \Big|_{y_{ij}=x_{ij}\sqrt{3}}. \tag{3.95}$$

Definition 3.19. The concurrence of 2-qutrit state (3.95) can informally be treated as the space diagonal of the rectangular parallelepiped (cuboid) built on three ancilla qutrit cross product vectors $| \Phi_i\rangle \times \left| \Phi_j \right\rangle$, $(i, j) = (0, 1), (1, 2), (2, 0)$ with the further substitution (3.91). We call this procedure a cross product concurrence computation.

Innovation 3.20. (Cross product of 7-qudits). *The cross product of two vectors (without modifications or extensions of its standard definition) exists in three and seven dimensions only* (Brown and Gray 1967). *Therefore, the above general method of concurrence construction can be provided in a similar way for two 7-qudit state using 7 ancilla 7-qudit cross products. The final formula* $C_7\left(| \Psi(2)\rangle_{(7)}\right)$ *will have possibly the same shape as that of 2-qutrit state (3.95), but with 7 summands.*

Innovation 3.21. (Cross product of n-qudits). *The cross product can be defined in n dimensions if we wish to modify it by an additional cross term* (Silagadze 2002, Tian et al 2013). *In the same way, the concurrence for n-qudit state* $C_n\left(| \Psi(2)\rangle_{(n)}\right)$ *can be computed through their cross products using the above procedure of introducing n ancilla n-qudits (3.87) and considering the space diagonal of the rectangular parallelepiped in n dimensions, similarly to (3.95).*

Innovation 3.22. (Cross product of 7-superqudits). *The cross product concurrence computation procedure can also be applied for superqubits because they are effectively* *defined in three-dimensional space (superspace* $\mathscr{H}_{(2|1)}$*), as well as for 7-superqubits in effective seven-dimensional superspace* $\mathscr{H}_{(r|s)}$, $r + s = 7$*. We can introduce a super-analog of (3.63), at least informally, taking into account* $\mathbb{Z}_2$ *sign rule in further calculations and use the graded involution (3.4) instead of the complex conjugation.*

References

Agarwal S and Wang X 2010 Qutrit influence on entanglement dynamics *Frontiers in Optics 2010/ Laser Science XXVI: OSA*

Ahadpour S and Mirmasoudi F 2018 Thermal quantum discord and super quantum discord teleportation via a two-qubit spin-squeezing model *Theor. Math. Phys.* **195** 628–39

Bagchi B K 2001 *Supersymmetry in Quantum and Classical Mechanics vol 116 of Monographs and Surveys in Pure and Applied Mathematics* (Chapman & Hall: CRC)

Baker J M, Duckering C, Gokhale P, Brown N C, Brown K R and Chong F T 2020 Improved quantum circuits via intermediate qutrits *ACM Transactions on Quantum Computing* **1** 1–25

Belhaj A, Sedra M B and Segui A 2014 Graph theory and qubit information systems of extremal black branes *J. Phys. A: Math. Theor.* **48** 045401

Bennett C H, DiVincenzo D P, Smolin J A and Wootters W K 1996 Mixed-state entanglement and quantum error correction *Phys. Rev.* **A54** 3824–51

Berezin F A 1987 *Introduction to Superanalysis* (Dordrecht: Reidel)

Berezin F A and Leites D A 1975 Supermanifolds *Soviet Math. Dokl* **16** 1218–22

Bernstein J, Leites D, Molotkov V and Shander V 2013 *Seminars of Supersymmetries. vol.1. Algebra and calculus* (Moscow: MCCME) In Russian, the English version is available for perusal

Binicioglu S, Klyachko A A and Shumovsky A S 2007 Single qutrit entanglement *Conference on Coherence and Quantum Optics: OSA CMI50*

Borsten L, Brádler K and Duff M J 2014 Tsirelson's bound and supersymmetric entangled states *Proc. Royal Soc. A* 470 0253

Borsten L, Brádler K and Duff M J 2015 The structure of superqubit states *Rev. Nuovo Cim.* **38** 371–86

Borsten L, Dahanayake D, Duff M J and Rubens W 2010 Superqubits *Phys. Rev.* **D81** 105023

Borsten L, Duff M J and Rubens W 2012 Black holes and qubits *Frontiers In Quantum Information Research* ed N Mikio (Singapure: World Scientific Publishing Co)

Brádler K 2012 *The Theory of Superqubits—Supersymmetric Qubits* 39 (arxiv:1208.2978)

Brown R B and Gray A 1967 Vector cross products *Comment. Math. Helv.* **42** 222–36

Castellani L, Grassi P A and Sommovigo L 2010 *Quantum Computing with Superqubits* preprint INFN, Univ. Piemonte Alessandria 13 p (arxiv:hep-th/1001.3753)

Caves C M and Milburn G J 2000 Qutrit entanglement *Opt. Commun.* **179** 439–46

Cerchiai B L and van Geemen B 2010 From qubits to e7 *J. Math. Phys.* **51** 122203

Cereceda J L 2003 *Degree of Entanglement for Two Qutrits in a Pure State* preprint arXiv Madrid 8 p (arxiv:quant-ph/0305043)

Constantinescu F 2002 Supersymmetric positivity and supersymmetric Hilbert space *Lett. Math. Phys.* **62** 111–25

Cooper F, Khare A and Sukhatme U 2001 *Supersymmetry in Quantum Mechanics* (River Edge: World Sci. Publishing Co.)

Dahanayake D 2010 The role of supersymmetry in the black hole/qubit correspondence: http:// arXiv.org/abs/Imperial Coll. London 178 p

Dehghani A, Akhound A and Panahyazdan F 2022 Crossed-product entangled states *Rep. Math. Phys.* **90** 257–70

Dehghani A, Mojaveri B and Alenabi A A 2021 Vector product approach of producing non-gaussian states *Int. J. Theor. Phys.* **60** 3885–95

Dehghani A, Mojaveri B and Bahrbeig R J 2021 Two-qutrit entangled f-coherent states *Rep. Math. Phys.* **87** 111–27

Deligne P, Lehrer G and Zhang R 2018 The first fundamental theorem of invariant theory for the orthosymplectic super group *Adv. Math.* **327** 4–24

De Witt B S 1992 *Supermanifolds* 2nd edn (Cambridge: Cambridge University Press)

Dirac P A M 1939 A new notation for quantum mechanics *Math. Proc. Cambridge Phil. Soc.* **35** 416–8

Duplij E S, Siegel W and J Bagger J 2004 *Concise Encyclopedia of Supersymmetry And Noncommutative Structures In Mathematics And Physics* (Dordrecht-Boston-London: Kluwer Academic Publishers) Second printing, Springer Science and Business Media, Berlin-New York-Heidelberg, 2005

Durkin G A and Simon C 2005 Multi-partite entanglement inequalities via spin vector geometry *Phys. Rev. Lett.* **95** 180402

Efetov K 1997 *Supersymmetry in Disorder and Chaos* (Cambridge: Cambridge University Press)

Eltschka C and Siewert J 2014 Quantifying entanglement resources *J. Phys. A: Math. Theor.* **47** 424005

Erol V 2017 *Analysis of Negativity and Relative Entropy of Entanglement Measures for Two Qutrit Quantum Communication Systems* (Istanbul: Okan University) 2017030217 MDPI preprint

Frydryszak A M 2013 Qubits, superqubits and squbits *J. Phys. Conf. Ser.* 411 012015

Gieres F 2000 Mathematical surprises and Dirac's formalism in quantum mechanics *Rep. Prog. Phys.* **63** 1893–931

Haba Z and Kupsch J 1995 Supersymmetry in Euclidean quantum field theory *Fortschr. Phys.* **43** 41–66

Hill S A and Wootters W K 1997 Entanglement of a pair of quantum bits *Phys. Rev. Lett.* **78** 5022–5

Holweck F and Jaffali H 2016 Three-qutrit entanglement and simple singularities *J. Phys. A: Math. Theor.* **49** 11

Horodecki R, Horodecki P, Horodecki M and Horodecki K 2009 Quantum entanglement *Rev. Mod. Phys.* **81** 865–942

Hruby J 2004 *Supersymmetry and Qubit Field Theory* preprint Institute of Physics Prague 20 p (quant-ph/0402188)

Hruby J 2008 *Super Quantum Mechanics, Spacetime and Quantum Information* (arxiv:quant-ph/0804.1442)

Huerta-Morales J D 2017 Entanglement in qutrit systems *Res. Comp. Sci.* **131** 9–13

Inoue R, Yonehara T, Miyamoto Y, Koashi M and Kozuma M 2009 Measuring qutrit-qutrit entanglement of orbital angular momentum states of an atomic ensemble and a photon *Phys. Rev. Lett.* **103** 110503

Junker G 1996 *Supersymmetric Methods in Quantum and Statistical Physics* (Berlin: Springer)

Khemakhmia A and Mebarki N 2021 On the super quantum information *J. Phys. Conf. Ser.* 1766 012032

Leites D and Serganova V 1991 Models of representations of some classical supergroups *Math. Scandinavica* **68** 131–47

Leites D A 1980 Introduction to the theory of supermanifolds *Russian Math. Surv.* **35** 1–64

Li Q, Cui J, Wang S and Long G-L 2017 Entanglement monogamy in three qutrit systems *Sci. Rep.* **7** 1946

Li W-A 2011 Distributed qutrit-qutrit entanglement via quantum zeno dynamics *Opt. Commun.* **284** 2245–9

Machnikowski P, Axt V M and Kuhn T 2007 Quantum-information encoding in dressed qubits *Phys. Rev.* A **75** 052330

Maleki Y 2016 Multi-mode entangled states represented as grassmannian polynomials *Quantum Inf. Process.* **15** 3893–907

Massey W S 1983 Cross products of vectors in higher dimensional Euclidean spaces *Amer. Math. Monthly* **90** 697–701

Mirzaei S, Najarbashi G, Fasihi M A and Mirmasoudi F 2018 Entanglement of multipartite fermionic coherent states for pseudo-hermitian Hamiltonians *Theor. Math. Phys.* **196** 1028–42

Mohapatra R N 1986 *Unification and Supersymmetry: the Frontiers of Quark-lepton Physics* (Berlin: Springer)

Pakhomov V F 1974 Automorphisms of tensor product of Abelian and Grassmann algebras *Mat. Zametki* **16** 65–74

Pashaev O K 2023 Maximally entangled two-qutrit quantum information states and de Gua's theorem for tetrahedron *Mathematical Methods for Engineering Applications vol 414 of Springer Proc. Math. Stat.* ed F Q-D Yilmaz, A M Vaquero and J Mierluş-Mazilu (Cham: World Scientific Publishing Co) 93–104 pp

Popescu S and Rohrlich D 1994 Quantum nonlocality as an axiom *Found. Phys.* **24** 379–85

Rogers A 1980 A global theory of supermanifolds *J. Math. Phys.* **21** 1352–65

Rogers A 2007 *Supermanifolds. Theory and Applications* (Singapore: World Scientific)

Rudin W 1991 *Functional Analysis* 2nd edn (New York: McGraw-Hill)

Rudolph O 2000 Super hilbert spaces *Commun. Math. Phys.* **214** 449–67

Silagadze Z K 2002 Multi-dimensional vector product *J. Phys. A: Math. Gen.* **35** 4949–53

Slavnov D A 2015 Computer model of a qubit *Phys. Particles Nuclei Lett.* **12** 439–42

Li H and yu Chen X 2011 Entanglement of graph qutrit states *2011 Int. Conf. Intelligence Science and Information Engineering* (NJ: IEEE) 61–4 pp

Terning J 2005 *Modern Supersymmetry* (Oxford: Oxford Univ. Press)

Tian X-L, Yang C, Hu Y and Tian C 2013 *Vector Cross Product in* n*-Dimensional Vector Space Mountain View* 9 p (arxiv:1310.5197)

van Eijndhoven S J L and de Graaf J 1985 A mathematical interpretation of Dirac's formalism for quantum mechanics *Lecture Notes in Math.* 209–51

van Nieuwenhuizen P and West P 1989 *Principles of Supersymmetry and Supergravity* (Cambridge: Cambridge University Press)

Vedral V, Plenio M B, Rippin M A and Knight P L 1997 Quantifying entanglement *Phys. Rev. Lett.* **78** 2275–9

Verstraete F, Dehaene J and DeMoor B 2001 Local filtering operations on two qubits *Phys. Rev.* **A64** 010101

Vidal G 2000 Entanglement monotones *J. Mod. Opt* **47** 355–76

Wang M, Xia Y, Li Y, Yang Y, Cao L, Zhang Q and Zhao J 2020 Protecting qutrit-qutrit entanglement under decoherence via weak measurement and measurement reversal *Int. J. Theor. Phys.* **59** 3696–704

Wegner F 2016a Path integrals for fermions and bosons *Supermathematics and its Applications in Statistical Physics* ed F Wegner (Berlin-Heidelberg: Springer) 47–66 pp

Wegner F 2016b *Supermathematics and its applications in statistical physics Grassmann variables and the method of supersymmetry vol 920 of Lecture Notes in Physics* (Berlin: Springer)

Wess J and Bagger J 1983 *Supersymmetry and Supergravity* (Princeton: Princeton University Press)

Wootters W K 1998 Entanglement of formation of an arbitrary state of two qubits *Phys. Rev. Lett.* **80** 2245–8

Xiao X and Li Y-L 2013 Protecting qutrit-qutrit entanglement by weak measurement and reversal *Eur. Phys. J.* D **67** 204

Chapter 4

Duality quantum computing

The duality (quantum) computer is based on the interference principle of any quantum system, but in a special way (Long 2006a, Long and Liu 2008). The main idea is to consider an undisturbed quantum system from the wave viewpoint, while on the measurement stage it is treated from the particle viewpoint. In this way, the initial quantum state (as wave) can be (1) decomposed into subwaves moving along separate paths and (2) combined at some point where they interfere (Gudder 2007). These two operations provide the additional duality parallelism, which can improve the calculational characteristics and the possible superiority of a duality computer (Long 2006a, Gudder 2008). The corresponding two additional operations are quantum operators of a new kind (duality gates): (quantum wave) divider and (quantum wave) combiner. The subwaves pass through a set of quantum gates and are collected by the combiner. The measurement is then performed on the joint final state. This procedure is a division of state of the same particle but is not a clone of the state of one particle onto another particle, and therefore this does not violate the no cloning theorem (Long 2011). The connection of the duality computer concept with the interference principle and computational applications was given in Long and Liu (2008), and the experimental realization was given in Wei *et al* (2017). Here we outline general mathematical constructions of a duality computer and present a new interpretation based on analogy with a convolution product in the polyadic Hopf algebra theory (Duplij 2022), which can be interesting by itself.

4.1 Duality computing and polyadic operations

Let us consider the complex Hilbert space $\mathcal{H}$ with the inner product $\langle\ |\rangle$ and denote the direct sum of n its copies by $\mathcal{H}^{\oplus n} = \oplus_{i=1}^{n}\mathcal{H}_i = \overbrace{\mathcal{H}_1 \oplus \cdots \oplus \mathcal{H}_n}^{n}$, where $\mathcal{H}_i = \mathcal{H}$, $1 \leqslant i \leqslant n$. For short, we also use the vector-like notation $\overrightarrow{\mathcal{H}} = \mathcal{H}^{\oplus n}$, such that the total quantum state $\oplus_{i=1}^{n}|\,\psi_i\rangle$ becomes

$$\left|\vec{\Psi}\right\rangle = \left|\vec{\Psi}(n)\right\rangle = \begin{pmatrix} |\psi_1\rangle \\ |\psi_2\rangle \\ \vdots \\ |\psi_n\rangle \end{pmatrix} \in \vec{\mathcal{H}}, \quad |\psi_i\rangle \in \mathcal{H}_i, \quad 1 \leqslant i \leqslant n, \tag{4.1}$$

where we use number of slits n in the vector state (or operators below) manifestly when it will be needed.

If all the states in (4.1) are the same $|\psi_i\rangle = |\psi\rangle$, then we place the subscript (=) as follows $\left|\vec{\Psi}\right\rangle = \left|\vec{\Psi}_=\right\rangle$, and this state will be called symmetric. A similar brief notation will be used for other variables taking values in the direct sum.

The total inner product $\langle\ |\rangle^{\rightarrow}$ of two vectors $\left|\vec{\Psi}\right\rangle$ and $\left|\vec{\Phi}\right\rangle$ is defined by (in the bra-ket notation)

$$\left\langle\vec{\Phi}\middle|\vec{\Psi}\right\rangle^{\rightarrow} = \langle\varphi_1|\psi_1\rangle + \langle\varphi_2|\psi_2\rangle + \cdots + \langle\varphi_n|\psi_n\rangle \in \mathbb{C}, \quad |\varphi_i\rangle, |\psi_i\rangle \in \mathcal{H}_i = \mathcal{H}, \quad 1 \leqslant i \leqslant n. \tag{4.2}$$

The space $\vec{\mathcal{H}}$ endowed with the total inner product $\langle\ |\rangle^{\rightarrow}$ (4.2) becomes a complex Hilbert space. The norm of the total space $\|\ \|^{\rightarrow}$ is induced by (4.2)

$$\left\|\vec{\Psi}\right\|^{\rightarrow} = \sqrt{\|\psi_1\|^2 + \|\psi_2\|^2 + \cdots + \|\psi_n\|^2}, \tag{4.3}$$

where $\|\psi_i\| = \sqrt{\langle\psi_i|\psi_i\rangle}$ is the norm in $\mathcal{H}_i$.

Thus, we have four different mappings of Hilbert spaces $\mathcal{H}$ and $\vec{\mathcal{H}}$

$$\mathcal{H} \to \mathcal{H}, \tag{4.4}$$

$$\vec{\mathcal{H}} \to \vec{\mathcal{H}}, \tag{4.5}$$

$$\mathcal{H} \to \vec{\mathcal{H}}, \tag{4.6}$$

$$\vec{\mathcal{H}} \to \mathcal{H}. \tag{4.7}$$

The first two mappings (4.4)–(4.5) are the standard (bounded linear) operators in the Hilbert spaces, such that $\mathbf{T}$: $\mathcal{H} \to \mathcal{H}$ and $\vec{\mathbf{T}}: \vec{\mathcal{H}} \to \vec{\mathcal{H}}$, where

$$\vec{\mathbf{T}} = \vec{\mathbf{T}}(n) = \begin{pmatrix} \mathbf{T}_1 \\ \mathbf{T}_2 \\ \vdots \\ \mathbf{T}_n \end{pmatrix}, \quad \mathbf{T}_i\colon \mathcal{H} \to \mathcal{H}, \tag{4.8}$$

The action on the total quantum state becomes

$$\vec{\mathbf{T}}\left|\vec{\Psi}\right\rangle = \begin{pmatrix} \mathbf{T}_1 \\ \mathbf{T}_2 \\ \vdots \\ \mathbf{T}_n \end{pmatrix} \bullet \begin{pmatrix} |\psi_1\rangle \\ |\psi_2\rangle \\ \vdots \\ |\psi_n\rangle \end{pmatrix} = \begin{pmatrix} \mathbf{T}_1|\psi_1\rangle \\ \mathbf{T}_2|\psi_2\rangle \\ \vdots \\ \mathbf{T}_n|\psi_n\rangle \end{pmatrix}, \tag{4.9}$$

and we informally define $\mathbf{T}_i \mid \psi_i\rangle := \mid \mathbf{T}_i\, \psi_i\rangle$, such that $\vec{\mathbf{T}} \left| \vec{\Psi} \right\rangle := \left| \vec{\mathbf{T}}\, \vec{\Psi} \right\rangle$, and $(\bullet)$ is the Hadamard product (here it is the componentwise action). The norm of the total operator $\vec{\mathbf{T}}$ is defined by analogy with (4.3)

$$\left\| \vec{\mathbf{T}} \right\| = \sqrt{\| \mathbf{T}_1 \|^2 + \| \mathbf{T}_2 \|^2 + \cdots + \| \mathbf{T}_n \|^2}, \tag{4.10}$$

where $\| \mathbf{T}_i \| = \sup \left\{ \frac{\| \mathbf{T}_i \psi_i \|}{\| \psi_i \|}, \forall \mid \psi_i\rangle \in \mathcal{H}_i, \mid \psi_i\rangle \neq 0 \right\}$.

The inner product in $\mathcal{H}$ of any quantum state $\mid \varphi\rangle \in \mathcal{H}$ with the transformed state $\mathbf{T} \mid \psi\rangle$ can be written as the functional $((\langle \varphi \mid)(\mathbf{T} \mid \psi\rangle)) = \langle \varphi | \mathbf{T}\, \psi\rangle = \langle \varphi \mid \mathbf{T} \mid \psi\rangle$, which is the convolution (in $\mathbb{C}$) of the operator $\mathbf{T}$ with the states $\mid \varphi\rangle$ and $\mid \psi\rangle$. The convolution of the operator $\vec{\mathbf{T}}$ in the total Hilbert space $\vec{\mathcal{H}}$ should be written with respect to the total inner product $\langle \mid \rangle^{\rightarrow}$ (4.2) in the following way

$$\left\langle \vec{\Phi} \middle| \vec{\mathbf{T}} \middle| \vec{\Psi} \right\rangle^{\rightarrow} = \langle \varphi_1 \mid \mathbf{T}_1 \mid \psi_1\rangle + \langle \varphi_2 \mid \mathbf{T}_2 \mid \psi_2\rangle + \cdots + \langle \varphi_n \mid \mathbf{T}_n \mid \psi_n\rangle \in \mathbb{C}. \tag{4.11}$$

If the states are the same $\mid \varphi\rangle = \mid \psi\rangle$, then the operator convolution $\langle \psi \mid \mathbf{T} \mid \psi\rangle$ is called the expectation value of the operator $\mathbf{T}$, and the total expectation value of $\vec{\mathbf{T}} \in \vec{\mathcal{H}}$ is determined using (4.11) with $\left| \vec{\Phi} \right\rangle = \left| \vec{\Psi} \right\rangle$.

The adjoint operator $\mathbf{T}^*$ with respect to the inner product $\langle \mid \rangle$ in $\mathcal{H}$ is defined by

$$\langle \varphi \mid \mathbf{T}^* \mid \psi\rangle = \overline{\langle \psi \mid \mathbf{T} \mid \varphi\rangle} \in \mathbb{C}, \tag{4.12}$$

where $\overline{()}$ is the complex conjugation. The corresponding adjoint operator $\overrightarrow{\mathbf{T}^*}$ in the total Hilbert space $\vec{\mathcal{H}}$ is defined in the similar way with respect to the total inner product

$$\left\langle \vec{\Phi} \middle| \vec{\mathbf{T}}^* \middle| \vec{\Psi} \right\rangle^{\rightarrow} = \overline{\left\langle \vec{\Psi} \middle| \mathbf{T} \middle| \vec{\Phi} \right\rangle^{\rightarrow}}, \tag{4.13}$$

which can be written using (4.11) as

$$\left\langle \vec{\Phi} \middle| \vec{\mathbf{T}}^* \middle| \vec{\Psi} \right\rangle^{\rightarrow} = \overline{\langle \psi_1 \mid \mathbf{T}_1 \mid \varphi_1\rangle} + \overline{\langle \psi_2 \mid \mathbf{T}_2 \mid \varphi_2\rangle} + \cdots + \overline{\langle \psi_n \mid \mathbf{T}_n \mid \varphi_n\rangle} \in \mathbb{C}. \tag{4.14}$$

The convolution can be written as $\langle \varphi \mid \mathbf{T} \mid \psi\rangle = ((\langle \varphi \mid \mathbf{T})(\mid \psi\rangle))$, where the bra vector $((\langle \varphi \mid \mathbf{T})$ in the matrix notation corresponds to $\varphi^\dagger \mathbf{T}$, which is equal to $(\mathbf{T}^\dagger \varphi)^\dagger$ (because $\mathbf{T}^* \longrightarrow \mathbf{T}^\dagger$ in the matrix notation), where $(\dagger)$ is the Hermitean conjugation. Thus, informally we can define $\langle \varphi \mid \mathbf{T} := \langle \mathbf{T}^*\, \varphi \mid$ to get the conventional relation for the adjoint operator

$$\langle \varphi | \mathbf{T}\, \psi\rangle = \langle \mathbf{T}^*\, \varphi | \psi\rangle. \tag{4.15}$$

In the total space we informally define by analogy $\left\langle \vec{\Phi} \right| \vec{\mathbf{T}} := \left\langle \vec{\mathbf{T}}^*\, \vec{\Phi} \right|$, and we have with respect to the total inner product in the total Hilbert space $\vec{\mathcal{H}}$

$$\left\langle \vec{\Phi} \middle| \vec{\mathbf{T}}\, \vec{\Psi} \right\rangle^{\rightarrow} = \left\langle \vec{\mathbf{T}}^{*}\, \vec{\Phi} \middle| \vec{\Psi} \right\rangle^{\rightarrow} \in \mathbb{C}, \tag{4.16}$$

which can be expanded using (4.11)–(4.14) to obtain

$$\left\langle \varphi_1 \middle| \mathbf{T}_1\, \psi_1 \right\rangle + \left\langle \varphi_2 \middle| \mathbf{T}_2\, \psi_2 \right\rangle + \cdots + \left\langle \varphi_n \middle| \mathbf{T}_n\, \psi_n \right\rangle = \left\langle \mathbf{T}_1^*\, \varphi_1 \middle| \psi_1 \right\rangle + \left\langle \mathbf{T}_2^*\, \varphi_2 \middle| \psi_2 \right\rangle + \cdots + \left\langle \mathbf{T}_n^*\, \varphi_n \middle| \psi_n \right\rangle. \tag{4.17}$$

It follows from (4.11), (4.17), and the commutativity of $\mathbb{C}$ that knowing the operator convolutions (which are in $\mathbb{C}$) in each subspace determines the total convolutions uniquely, but not vice versa.

Recall that the unitary operator $\mathbf{T} = \mathbf{U}$ preserves the inner product in the Hilbert space

$$\langle \mathbf{U}\, \varphi | \mathbf{U}\, \psi \rangle = \langle \varphi | \psi \rangle \in \mathbb{C}. \tag{4.18}$$

Using (4.15), we standardly obtain that unitary operators satisfy

$$\mathbf{U}^* \circ \mathbf{U} = \mathbf{U} \circ \mathbf{U}^* = \mathrm{id}. \tag{4.19}$$

In the direct sum of spaces $\vec{\mathcal{H}}$ we have the definition of $\vec{\mathbf{U}}$ with the respect of the total inner product

$$\left\langle \vec{\mathbf{U}}\, \vec{\Phi} \middle| \vec{\mathbf{U}}\, \vec{\Psi} \right\rangle^{\rightarrow} = \left\langle \vec{\Phi} \middle| \vec{\Psi} \right\rangle \in \mathbb{C}, \tag{4.20}$$

and by means of (4.16) we obtain

$$\vec{\mathbf{U}}^{*} \circ \vec{\mathbf{U}} = \vec{\mathbf{U}} \circ \vec{\mathbf{U}}^{*} = \vec{\mathrm{id}}, \quad \vec{\mathrm{id}} = \overbrace{\mathrm{id} \oplus \mathrm{id} \oplus \cdots \oplus \mathrm{id}}^{n}. \tag{4.21}$$

The unitary operators $\{\mathbf{U}\}$ acting in $\mathcal{H}$ are widely used as quantum gates in quantum computers, while the vector operators $\{\vec{\mathbf{U}}\}$ act in the total space $\vec{\mathcal{H}}$ and are exploited as vector quantum gates in duality computing (Long and Liu 2008).

The second two mappings (4.6)–(4.7) can have another meaning (than the previous ordinary operators acting in some Hilbert space) because they are multiary operations between Hilbert spaces: $\mathcal{H}$ and $\vec{\mathcal{H}}$. We propose to treat them as n-ary comultiplication (4.6) and n-ary multiplication (4.7) of the special kind.

Definition 4.1. Let $|\,\psi\rangle \in \mathcal{H}$ be a quantum state in the Hilbert space $\mathcal{H}$, then the divider $\mathbf{D}_{\mathrm{p}}$ is the operation which splits the quantum wave into n weighted subwaves, describing multi-slits, as

$$\mathbf{D}_{\mathrm{p}}(n)\colon \mathcal{H} \to \vec{\mathcal{H}} = \mathcal{H}^{\oplus n}, \tag{4.22}$$

$$\mathbf{D}_{\mathrm{p}}(n)\,|\,\psi\rangle = \begin{pmatrix} p_1 \\ p_2 \\ \vdots \\ p_n \end{pmatrix} \bullet \begin{pmatrix} |\,\psi\rangle \\ |\,\psi\rangle \\ \vdots \\ |\,\psi\rangle \end{pmatrix} = \begin{pmatrix} p_1\,|\,\psi\rangle \\ p_2\,|\,\psi\rangle \\ \vdots \\ p_n\,|\,\psi\rangle \end{pmatrix} = \vec{P} \bullet \left| \vec{\Psi}_{=} \right\rangle, \quad p_i \in \mathbb{C}, \tag{4.23}$$

where the probability distribution $\{\vec{P}\}$ is called the divider structure. The divider structure is normalized, if

$$\left(\vec{P}\right)^{\dagger}\vec{P} = (\bar{p}_1, \bar{p}_2, \ldots, \bar{p}_n)\begin{pmatrix} p_1 \\ p_2 \\ \vdots \\ p_n \end{pmatrix} = \sum_{i=1}^{n} | p_i |^2 = 1, \quad p_i \in \mathbb{C}. \tag{4.24}$$

Definition 4.2. The divider structure is called uniform, if

$$p_1 = p_2 = \cdots = p_n = \frac{1}{n}. \tag{4.25}$$

The divider operation $\mathbf{D}_{\mathrm{p}}(n)$ (4.22) can be treated a special analog of n-ary comultiplication (deformed n-ary coaddi) map ($\Delta^{(n)}$) in the polyadic Hopf algebra theory (see Duplij 2022, chapter 9). The analog of polyadic total coassociativity for $\mathbf{D}_{\mathrm{p}}(n)$ is

$$\begin{gathered}\left(\overbrace{\mathrm{id} \oplus \cdots \oplus \mathrm{id}}^{n-1-i} \oplus \mathbf{D}_{\mathrm{p}}(n) \oplus \overbrace{\mathrm{id} \oplus \cdots \oplus \mathrm{id}}^{i}\right) \circ \mathbf{D}_{\mathrm{p}}(n) \\ = \left(\overbrace{\mathrm{id} \oplus \cdots \oplus \mathrm{id}}^{n-1-j} \oplus \mathbf{D}_{\mathrm{p}}(n) \oplus \overbrace{\mathrm{id} \oplus \cdots \oplus \mathrm{id}}^{j}\right) \circ \mathbf{D}_{\mathrm{p}}(n), \\ \forall\, i, j = 0, \ldots, n-1, \ i \neq j, \ \mathrm{id} \equiv \mathrm{id}_{\mathrm{H}}: \mathcal{H} \to \mathcal{H}, \ \mathrm{id} \circ | \psi \rangle = | \psi \rangle. \end{gathered} \tag{4.26}$$

There are $(n-1)^2$ relations in (4.26). If not all of them satisfied the polyadic coassociativity, then this is called partial (see, e.g. Thurston 1949, Belousov 1972, Sokhatsky 1997).

Definition 4.3. Let $| \vec{\Phi} \rangle \in \vec{\mathcal{H}}$ be a direct sum of quantum states (4.1), then the combiner operation that gathers multi-slits in one quantum state as follows

$$\mathbf{C}_{\mathrm{q}}(n): \vec{\mathcal{H}} \to \mathcal{H}, \tag{4.27}$$

$$\mathbf{C}_{\mathrm{q}}(n) \left| \vec{\Psi} \right\rangle = \mathbf{C}_{\mathrm{q}} \begin{pmatrix} | \psi_1 \rangle \\ | \psi_2 \rangle \\ \vdots \\ | \psi_n \rangle \end{pmatrix} = q_1 | \psi_1 \rangle + q_2 | \psi_1 \rangle + \cdots + q_n | \psi_n \rangle = \vec{Q}^{T} \left| \vec{\Psi} \right\rangle \in \mathcal{H}, \quad q_i \in \mathbb{C}, \tag{4.28}$$

where the probability distribution $\vec{Q}^{T} = \{q_1, q_2, \ldots, q_n\}$ is called the combiner structure, and it is normalized, if

$$\left(\vec{Q}\right)^T \vec{Q} = (q_1, q_2, \ldots, q_n) \begin{pmatrix} \bar{q}_1 \\ \bar{q}_2 \\ \vdots \\ \bar{q}_n \end{pmatrix} = \sum_{i=1}^{n} |q_i|^2 = 1, \quad q_i \in \mathbb{C}. \tag{4.29}$$

The combiner operation $\mathbf{C}_{\mathrm{q}}(n)$ (4.27) can be treated as a deformed (by the probability distribution $\vec{Q}$) analog of n-ary multiplication (or more exactly addition) map ($\mu^{(n)}$) in the polyadic algebra theory (Duplij 2022, chapter 5). The analog of polyadic total associativity for $\mathbf{C}_{\mathrm{q}}$ is

$$\begin{aligned} &\mathbf{C}_{\mathrm{q}}(n) \circ \left(\overbrace{\mathrm{id} \oplus \cdots \oplus \mathrm{id}}^{n-1-i} \oplus \mathbf{C}_{\mathrm{q}}(n) \oplus \overbrace{\mathrm{id} \oplus \cdots \oplus \mathrm{id}}^{i} \right) \\ &= \mathbf{C}_{\mathrm{q}}(n) \circ \left(\overbrace{\mathrm{id} \oplus \cdots \oplus \mathrm{id}}^{n-1-j} \oplus \mathbf{C}_{\mathrm{q}}(n) \oplus \overbrace{\mathrm{id} \oplus \cdots \oplus \mathrm{id}}^{j} \right), \\ &\forall i, j = 0, \ldots, n-1, \ i \neq j, \ \mathrm{id} \equiv \mathrm{id}_{\mathrm{H}}: \mathcal{H} \to \mathcal{H}, \ \mathrm{id} \circ |\psi\rangle = |\psi\rangle. \end{aligned} \tag{4.30}$$

There are $(n-1)^2$ relations in (4.26). If not all of them satisfied the polyadic associativity, then this is called partial (see, e.g. Thurston 1949, Belousov 1972).

Proposition 4.4. The combiner operation $\mathbf{C}_{\mathrm{q}}(n)$ is totally polyadic associative, if the probability distribution $\left\{\vec{Q}\right\}$ is idempotent

$$\vec{Q} \bullet \vec{Q} = \vec{Q}, \quad \text{or } q_i^2 = q_i, \quad i = 1, \ldots, n. \tag{4.31}$$

Proof. Make the action of both sides of (4.30) on $|\psi\rangle$ and insert (4.28) into each of n places consequently to get (4.31).

Corollary 4.5. Because $q_i \in \mathbb{C}$, and in $\mathbb{C}$ there only two idempotents, that are 0 and 1, the total associativity of $\mathbf{C}_{\mathrm{q}}$ for invertible q_i implies that all $q_i = 1$, $i = 1, \ldots, n$.

Corollary 4.6. If some q_i in (4.28) are not idempotent (4.31), then the combiner operation is a polyadic operator (n-ary multiplication) $\mathbf{C}_{\mathrm{q}}: \vec{\mathcal{H}} \to \mathcal{H}$ that is not totally associative.

Definition 4.7. The combiner structure $\left\{\vec{Q}\right\}$ is called uniform, if

$$q_1 = q_2 = \cdots = q_n = \frac{1}{n}. \tag{4.32}$$

The nonassociative binary operators are widely used in quantum mechanics and quantum field theory (Løhmus *et al* 1994).

Let us consider possible relations between divider and combiner operations. Initially, we will not fix the probability distributions (in our approach, deformation parameters) $\left\{\vec{P}\right\}$ and $\left\{\vec{Q}\right\}$, trying to find their connections in special cases.

Proposition 4.8. The composition ($\circ$) of divider and combiner is the identity operator in $\mathcal{H}$, if

$$\mathbf{C}_{\mathrm{q}}(n)\circ\mathbf{D}_{\mathrm{p}}(n) = \mathrm{id} \Longleftrightarrow q_1p_1 + q_2p_2 + \cdots + q_np_n = 1, \qquad q_i, p_i \in \mathbb{C}. \tag{4.33}$$

Proof. It follows directly from consequent acting first of $\mathbf{D}_{\mathrm{p}}$ and then $\mathbf{C}_{\mathrm{q}}$ on the quantum state $|\psi\rangle$ and then using the definitions (4.23) and (4.28).

Corollary 4.9. A particular case

$$q_i = \bar{p}_i \tag{4.34}$$

corresponds to the complex duality computing (Cao *et al* 2012), in this choice the condition (4.33) leads to uniformity of both $\mathbf{C}_{\mathrm{q}}(n)$ (4.32) and $\mathbf{D}_{\mathrm{p}}(n)$ (4.25).

Without the restrictions (4.34) the equation (4.33) has an infinite number of solutions, even when both divider and combiner operations are uniform.

Definition 4.10. We say that the divider and combiner are consistent if they are similar to n-ary coalgebra map and n-ary algebra map, respectively (Duplij 2022, chapter 9), i.e., $\mathbf{D}_{\mathrm{p}}(n)$ and $\mathbf{C}_{\mathrm{q}}(n)$ satisfy

$$\mathbf{C}_{\mathrm{q}}(n)\circ\left(\overbrace{\mathbf{D}_{\mathrm{p}}(n) \oplus \cdots \oplus \mathbf{D}_{\mathrm{p}}(n)}^{n}\right)\left|\vec{\Psi}\right\rangle = \mathbf{D}_{\mathrm{p}}(n)\circ\mathbf{C}_{\mathrm{q}}(n)\left|\vec{\Psi}\right\rangle. \tag{4.35}$$

In the Hopf algebra theory, the consistency condition is the part of polyadic bialgebra definition in terms of n-ary multiplication and n-ary comultiplication (Duplij 2022, chapter 9).

Proposition 4.11. If the total state is symmetric $\rangle\left|\vec{\Psi}\right\rangle = \left|\vec{\Psi}_{=}\right\rangle$ (4.1), then the divider and combiner are consistent, when the probability distributions are connected by n equations

$$q_i(p_1 + p_2 + \cdots + p_n) = p_i(q_1 + q_2 + \cdots + q_n), \quad i = 1,\ldots,n, \quad p_i, q_i \in \mathbb{C}. \tag{4.36}$$

Proof. It follows directly from the definitions (4.23), (4.28) and the consistency condition (4.36).

In case $n = 2$, we have only one condition $q_1p_2 = q_2p_1$.

Definition 4.12. We introduce an analog of the polyadic ith partial antipode $\mathbf{S}_i$ (Duplij 2022, chapter 9) by

$$\mathbf{C}_{\mathrm{q}}(n)\circ\left(\overbrace{\mathrm{id}\oplus\cdots\oplus\mathrm{id}}^{n-1-i}\oplus\mathbf{S}_i(n)\oplus\overbrace{\mathrm{id}\oplus\cdots\oplus\mathrm{id}}^{i}\right)\circ\mathbf{D}_{\mathrm{p}}(n)\mid\psi\rangle=\mid\psi\rangle,\quad\mid\psi\rangle\in\mathcal{H}. \tag{4.37}$$

The full antipode is defined as $\mathbf{S}_{\mathrm{pq}} = \mathbf{S}_i(n)$, $i = 1,\ldots,n$ if all partial antipodes are equal.

For ith partial antipode from the definitions (4.23), (4.28) and (4.37) we obtain

$$\mathbf{S}_i(n)=\frac{1}{q_ip_i}\left(1-\sum_{k=1,k\neq i}^{n}q_kp_k\right),\quad p_i,\,q_i\in\mathbb{C}. \tag{4.38}$$

To finally understand the general algebraic structure of the duality computing, we turn to further similarity with the polyadic Hopf algebra theory (Duplij 2022, chapter 9) and introduce an analog of the n-ary convolution for gates. Recall that the binary convolution product of two operators $\mathbf{T}_1$ and $\mathbf{T}_2$ in the bialgebra having multiplication μ and comultiplication Δ is defined by $\mu\circ(\mathbf{T}_1\otimes\mathbf{T}_2)\circ\Delta$ (Abe 1980, Sweedler 1969, Radford 2012).

Let $\vec{\mathbf{T}}$ be an operator (4.8) in the total Hilbert space $\vec{\mathcal{H}}$ (in the vector notation), $\mathbf{D}_{\mathrm{p}}$ and $\mathbf{C}_{\mathrm{q}}$ be the divider (4.23) and combiner (4.28).

Definition 4.13. The duality n-ary convolution of the vector operator $\vec{\mathbf{T}}$ is the composition

$$\mathbf{Q}_{\mathrm{dual}}^{(n),q,p}\left(\vec{\mathbf{T}}\right)=\mathbf{C}_{\mathrm{q}}(n)\circ\left(\overbrace{\mathbf{T}_1\oplus\mathbf{T}_2\oplus\cdots\oplus\mathbf{T}_n}^{n}\right)\circ\mathbf{D}_{\mathrm{p}}(n)=\mathbf{C}_{\mathrm{q}}(n)\circ\vec{\mathbf{T}}\circ\mathbf{D}_{\mathrm{p}}(n). \tag{4.39}$$

In this way, the antipode (4.37) can be treated as the polyadic inverse of the identity with respect to the n-ary convolution product (4.39)

Informally, the ordinary quantum computation process (on pure states) consists of:

(1) Preparation of the initial state $\mid\psi\rangle_{\mathrm{init}}$.

(2) Computation as consequent action on $\mid\psi\rangle$ with the set of k quantum gates being unitary operators $\{\mathbf{U}^{(1)},\mathbf{U}^{(2)},\ldots,\mathbf{U}^{(k)}\}$ by their composition to obtain the final (still not measured) quantum state

$$\mid\psi\rangle_{\mathrm{fin}}=\mathbf{U}^{(k)}\circ\mathbf{U}^{(k-1)}\circ\cdots\circ\mathbf{U}^{(2)}\circ\mathbf{U}^{(1)}\mid\psi\rangle_{\mathrm{init}}=\mathbf{U}\mid\psi\rangle_{\mathrm{init}}. \tag{4.40}$$

(3) Measurement $\mathbf{M}$: $\mid\psi\rangle_{\mathrm{fin}}\longrightarrow\mid\psi\rangle_{\mathrm{measured}}$.

In the duality computation processing, one changes *(2)* and replaces (4.40) with the more complicated set of gates acting in subspaces. Indeed, if in each ith subspace we have k unitary gates

$$\vec{\mathbf{U}} = \begin{pmatrix} \mathbf{U}_1 \\ \mathbf{U}_2 \\ \vdots \\ \mathbf{U}_n \end{pmatrix}, \ \mathbf{U}_i = \mathbf{U}_i^{(k)} \circ \mathbf{U}_i^{(k-1)} \circ \cdots \circ \mathbf{U}_i^{(2)} \circ \mathbf{U}_i^{(1)}, \quad \mathbf{U}_i \colon \mathcal{H} \to \mathcal{H}, \ \ i = 1, \ldots, n, \tag{4.41}$$

then we obtain in total kn unitary gates.

Definition 4.14. Duality computation with n sub-slits is defined by kn unitary gates, vector unitary gates $\vec{\mathbf{U}}$ (instead of the standard unitary gate $\mathbf{U}$ (4.40) composed from k unitary gates)

$$|\,\psi\rangle_{\text{fin}} = \mathbf{Q}_{\text{dual}}^{(n),q,p}\left(\vec{\mathbf{U}}\right) |\,\psi\rangle_{\text{init}}, \tag{4.42}$$

where the duality operator $\mathbf{Q}_{\text{dual}}^{(n),q,p}\left(\vec{\mathbf{U}}\right)$ is defined in (4.39).

Effectively, any operator connecting initial and final quantum states can be called a generalized quantum gate, but duality quantum gates are special nonunitary combinations of the given unitary operators.

Theorem 4.15. The duality quantum gate $\mathbf{Q}_{\text{dual}}^{(n),q,p}\left(\vec{\mathbf{U}}\right)$ is a nonunitary operator of the form

$$\mathbf{T}_{\text{dual}} = \mathbf{Q}_{\text{dual}}^{(n),q,p}\left(\vec{\mathbf{U}}\right) = q_1 p_1 \mathbf{U}_1 + q_2 p_2 \mathbf{U}_2 + \cdots + q_n p_n \mathbf{U}_n, \ \ q_i, p_i \in \mathbb{C}. \tag{4.43}$$

Proof. It follows directly from (4.42) the divider (4.23) and combiner (4.28).

Remark 4.16. The duality gate (4.43) is nonunitary, but with the special choice of probability distributions $\left\{\vec{Q}\right\}$ and $\left\{\vec{P}\right\}$ one could obtain the unitary $\mathbf{T}_{\text{dual}}$.

If the probability distributions satisfy (4.33), then the product $q_i p_i$ can be treated as the probability of the quantum wave to pass through the ith slit.

Corollary 4.17. If all components of $\vec{\mathbf{U}}$ are equal and (4.33) satisfied, then the duality quantum computer reduces to the ordinary quantum computer.

The case $q_i = \bar{p}_i$ was considered in Cao *et al* (2012), and such $\mathbf{T}_{\text{dual}}$ was called generalized duality quantum gate.

4.2 Higher duality computing

Now we propose another generalization of the duality computation (4.42), i.e., the higher duality one, by using higher powers of the divider (4.23) and combiner (4.28). Because they are polyadic (multiary) operations, to be consistent with arities and number of entries, we should use the polyadic powers for them (Duplij 2022). The main consequence of this would be the possibility of having different compositions of the divider $\mathbf{D}_{\mathrm{p}}(n_p)$ (4.23) and combiner $\mathbf{C}_{\mathrm{q}}(n_{\mathrm{q}})$ (4.28) with different arities $n_p \neq n_{\mathrm{q}}$, such that the number of slits remains to be equal to n.

Definition 4.18. The polyadic power ℓ_{p} of the divider $\mathbf{D}_{\mathrm{p}}(n_{\mathrm{p}})$ (4.23) as n_{p}-ary comultiplication (coaddition) is defined by the coiterated coaction (or ℓ_{p} compositions)

$$\left(\mathbf{D}_{\mathrm{p}}(n_{\mathrm{p}})\right)^{\circ \ell_{\mathrm{p}}} = \overbrace{\left(\left(\overbrace{\mathrm{id} \oplus \cdots \oplus \mathrm{id}}^{n_{\mathrm{p}}-1} \oplus \cdots \left(\overbrace{\mathrm{id} \oplus \cdots \oplus \mathrm{id}}^{n_{\mathrm{p}}-1} \oplus \mathbf{D}_{\mathrm{p}}(n_{\mathrm{p}})\right) \cdots \circ \mathbf{D}_{\mathrm{p}}(n_{\mathrm{p}})\right) \circ \mathbf{D}_{\mathrm{p}}(n_{\mathrm{p}})\right.}^{\ell_{\mathrm{p}}}, \tag{4.44}$$

$$\mathrm{id}\colon \mathcal{H} \to \mathcal{H}, \quad \mathbf{D}_{\mathrm{p}}(n_{\mathrm{p}})\colon \mathcal{H} \to \mathcal{H}^{\oplus n_{\mathrm{p}}},$$

and the arity n'_{q} of the composed operation (4.46) is equal to

$$n'_{\mathrm{p}} = \ell_{\mathrm{p}}\left(n_{\mathrm{p}} - 1\right) + 1. \tag{4.45}$$

Definition 4.19. The polyadic power ℓ_{q} of the combiner (4.28) as n_{q}-ary multiplication (addition) is defined by the iterated action (or ℓ_{q} compositions)

$$\left(\mathbf{C}_{\mathrm{q}}(n_{\mathrm{q}})\right)^{\circ \ell_{\mathrm{q}}} = \overbrace{\mathbf{C}_{\mathrm{q}}(n_{\mathrm{q}}) \circ \left(\mathbf{C}_{\mathrm{q}}(n_{\mathrm{q}}) \circ \cdots \left(\mathbf{C}_{\mathrm{q}}(n_{\mathrm{q}})\right.\right.}^{\ell_{\mathrm{q}}} \oplus \overbrace{\mathrm{id} \oplus \cdots \oplus \mathrm{id}}^{n_{\mathrm{q}}-1}\Big) \cdots \oplus \overbrace{\mathrm{id} \oplus \cdots \oplus \mathrm{id}}^{n_{\mathrm{q}}-1}\Big)\Bigg), \tag{4.46}$$

$$\mathrm{id}\colon \mathcal{H} \to \mathcal{H}, \quad \mathbf{C}_{\mathrm{q}}(n_{\mathrm{q}})\colon \mathcal{H}^{\oplus n_{\mathrm{q}}} \to \mathcal{H},$$

and the arity n'_{q} of the composed operation (4.46) is equal to

$$n'_{\mathrm{q}} = \ell_{\mathrm{q}}\left(n_{\mathrm{q}} - 1\right) + 1. \tag{4.47}$$

Note that the brackets in the polyadic powers (4.46) and (4.44) can be omitted if operations $\mathrm{C}_{\mathrm{q}}(n_{\mathrm{q}})$ and $\mathbf{D}_{\mathrm{p}}(n_{\mathrm{p}})$ are totally associative and coassociative, respectively.

Now, by analogy with (4.33), the composition of ℓ_{p} power of divider and ℓ_{q} power of combiner can be the identity operator in $\mathcal{H}$ but not for all divider and combiner arities n_{p} and n_{q}.

Proposition 4.20. The composition of ℓ_{p} power of divider and ℓ_{q} power of combiner can be the identity operator in $\mathcal{H}$ if their arities satisfy

$$\mathbf{C}_{\mathrm{q}}(n_{\mathrm{q}})^{\circ \ell_{\mathrm{q}}} \circ \mathbf{D}_{\mathrm{p}}(n_{\mathrm{p}})^{\circ \ell_{\mathrm{p}}} = \mathrm{id}, \tag{4.48}$$

$$\ell_p(n_p - 1) = \ell_q(n_q - 1). \tag{4.49}$$

Proof. It follows directly from the definitions (4.23) and (4.28) that the arities of the powers should coincide $n_p' = n_q'$ and the arity formulas (4.45), (4.47).

Equation (4.49) can be treated as so called quantization of arities, and therefore the analog of (4.43) depends of their concrete values.

Example 4.21. Let us consider the minimal case of nonbinary ($n_{p,q} > 2$) and unequal arities $n_p = 3$, $\ell_p = 3$ and $n_q = 4$, $\ell_q = 2$, and the total number of slits $\ell_p(n_p - 1) + 1 = \ell_q(n_q - 1) + 1 = 7$. Then for the values of the polyadic powers of divider (4.44) and combiner (4.46) we have manifestly using (4.27) and (4.22)

$$\mathbf{D}_p(3)^{\circ 3} \mid \psi\rangle = \begin{pmatrix} p_1 \mid \psi\rangle \\ p_2 \mid \psi\rangle \\ p_1p_3 \mid \psi\rangle \\ p_2p_3 \mid \psi\rangle \\ p_1p_3^2 \mid \psi\rangle \\ p_2p_3^2 \mid \psi\rangle \\ p_3^3 \mid \psi\rangle \end{pmatrix}, \tag{4.50}$$

$$\mathbf{C}_q(4)^{\circ 2} \left| \vec{\Psi}(7) \right\rangle = q_1(q_1 \mid \psi_1\rangle + q_2 \mid \psi_2\rangle + q_3 \mid \psi_3\rangle + q_4 \mid \psi_4\rangle) + q_2 \mid \psi_5\rangle + q_3 \mid \psi_6\rangle + q_4 \mid \psi_7\rangle. \tag{4.51}$$

The condition that the composition of the powers (4.50) and (4.51) to be the identity (4.48) gives the equation for the probability distributions $\left\{\vec{Q}(4)\right\} = \{q_1, q_2, q_3, q_4\}$ and $\left\{\vec{P}(3)\right\} = \{p_1, p_2, p_3\}$

$$q_1^2p_1 + q_1q_2p_2 + q_1q_3p_1p_3 + q_1q_4p_2p_3 + q_2p_1p_3^2 + q_3p_2p_3^2 + q_4p_3^3 = 1, \tag{4.52}$$

which is nonlinear in q_i, p_i and should be compared with the standard linear case of unity powers (4.33).

Let us introduce the higher analog of the duality n-ary operator $\mathbf{Q}_{\text{dual}}^{(n),q,p}$ (4.39).

Definition 4.22. The higher duality n-ary convolution of the vector operator $\vec{\mathbf{T}} \in \vec{\mathcal{H}}$ is the composition of the ℓ_p dividers (4.44) and ℓ_q combiners (4.46) which maps $\mathcal{H} \to \mathcal{H}$

$$\mathbf{Q}_{\text{hdual}, \ell_q, \ell_p}^{(n),q,p}\left(\vec{\mathbf{T}}\right) = \mathbf{C}_q^{\circ\ell_q}(n_q) \circ \left(\overbrace{\mathbf{T}_1 \oplus \mathbf{T}_2 \oplus \cdots \oplus \mathbf{T}_n}^{n} \right) \circ \mathbf{D}_p^{\circ\ell_p}(n_p) = \mathbf{C}_q^{\circ\ell_q}(n_q) \circ \vec{\mathbf{T}} \circ \mathbf{D}_p^{\circ\ell_p}(n_p), \tag{4.53}$$

where the number of slits n is equal to

$$n = \ell_{\mathrm{p}}(n_{\mathrm{p}} - 1) = \ell_{\mathrm{q}}(n_{\mathrm{q}} - 1). \tag{4.54}$$

Definition 4.23. The higher duality computation with n sub-slits is defined by kn unitary gates, vector unitary gates $\vec{\mathbf{U}}$ (4.41)

$$|\,\psi\rangle_{\mathrm{fin}} = \mathbf{Q}^{(n),\mathrm{q},\mathrm{p}}_{\mathrm{hdual},\,\ell_{\mathrm{q}},\,\ell_{\mathrm{p}}}\left(\vec{\mathbf{U}}\right)|\,\psi\rangle_{\mathrm{init}}, \tag{4.55}$$

where the duality operator $\mathbf{Q}^{(n),\mathrm{q},\mathrm{p}}_{\mathrm{hdual},\,\ell_{\mathrm{q}},\,\ell_{\mathrm{p}}}\left(\vec{\mathbf{U}}\right)$ is defined in (4.53).

Theorem 4.24. The $(\ell_{\mathrm{q}}, \ell_{\mathrm{p}})$-higher duality quantum gate $\mathbf{Q}^{(n),\mathrm{q},\mathrm{p}}_{\mathrm{hdual},\,\ell_{\mathrm{q}},\,\ell_{\mathrm{p}}}\left(\vec{\mathbf{U}}\right)$ is a non-unitary operator of the form

$$\mathbf{T}_{\mathrm{hdual}} = \mathbf{Q}^{(n),\mathrm{q},\mathrm{p}}_{\mathrm{hdual},\,\ell_{\mathrm{q}},\,\ell_{\mathrm{p}}}\left(\vec{\mathbf{U}}\right) = \mathbf{C}_{\mathrm{q}}^{\circ\ell_{\mathrm{q}}}(n)\circ\vec{\mathbf{U}}\circ\mathbf{D}_{\mathrm{p}}^{\circ\ell_{\mathrm{p}}}(n), \quad \mathbf{T}_{\mathrm{hdual}}\colon \mathcal{H} \to \mathcal{H}. \tag{4.56}$$

It is important that not all possible values of arities and powers are allowed, but only those which satisfy the quantization condition (4.54). The allowed number of slits n and corresponding n_{q}, n_{p} and ℓ_{q}, ℓ_{p} are presented in table 4.1. Note that the unusual peculiarity comes from the nondiagonal entries, which correspond to unequal arities of divider and combiner $n_{\mathrm{p}} \neq n_{\mathrm{q}}$. The table is symmetric, which means that the arity n (number of slits) is invariant under the exchange $(n_{\mathrm{p}}, \ell_{\mathrm{p}}) \longleftrightarrow (n_{\mathrm{q}}, \ell_{\mathrm{q}})$ following from (4.54).

Table 4.1. The allowed valued of slits for given arities n_{p}, n_{q} and polyadic powers (or numbers of divider and combiner compositions) ℓ_{p}, ℓ_{q}. The framed box corresponds to the binary standard duality convolution (4.39) with two slits $n = 2$.

n_q \ n_p	$\mathbf{C}_q$	$n_q = 2$			$n_q = 3$			$n_q = 4$			$n_q = 5$		
$\mathbf{D}_p$	ℓ_q \ ℓ_p	$\ell_q = 1$	$\ell_q = 2$	$\ell_q = 3$	$\ell_q = 1$	$\ell_q = 2$	$\ell_q = 3$	$\ell_q = 1$	$\ell_q = 2$	$\ell_q = 3$	$\ell_q = 1$	$\ell_q = 2$	$\ell_q = 3$
$n_p = 2$	$\ell_p = 1$	**2**											
	$\ell_p = 2$		**3**		**3**								
	$\ell_p = 3$			**4**				**4**					
$n_p = 3$	$\ell_p = 1$		**3**		**3**								
	$\ell_p = 2$					**5**					**5**		
	$\ell_p = 3$						**7**		**7**				
$n_p = 4$	$\ell_p = 1$			**4**				**4**					
	$\ell_p = 2$						**7**		**7**				
	$\ell_p = 3$									**10**			
$n_p = 5$	$\ell_p = 1$					**5**					**5**		
	$\ell_p = 2$											**9**	
	$\ell_p = 3$												**13**

Example 4.25. (*Example* 4.21 continued). With the concrete parameters (4.50) and (4.51), we have the 7-ary convolution product with three dividers and two combiners

$$\begin{aligned} \mathrm{T}_{\text{hdual}} &= \mathbf{Q}^{(n),\mathrm{q},\mathrm{p}}_{\text{hdual},\,\ell_{\mathrm{q}},\,\ell_{\mathrm{p}}}\left(\vec{\mathbf{U}}\right) = \mathbf{C}_{\mathrm{q}}(4)^{\circ 2}\circ\vec{\mathbf{U}}(7)\circ\mathbf{D}_{\mathrm{p}}(3)^{\circ 3}\mid\psi\rangle \\ &= q_1^2 p_1 \mathrm{U}_1 + q_1 q_2 p_2 \mathrm{U}_2 + q_1 q_3 p_1 p_3 \mathrm{U}_3 + q_1 q_4 p_2 p_3 \mathrm{U}_4 + q_2 p_1 p_3^2 \mathrm{U}_5 + q_3 p_2 p_3^2 \mathrm{U}_6 + q_4 p_3^3 \mathrm{U}_7. \end{aligned} \tag{4.57}$$

Effectively, a nonunitary operator connecting initial and final quantum states can be called a higher generalized quantum gate, i.e., higher duality quantum gate. Indeed, to continue *example* 4.21 and 4.25, and consider a nonunitary generalized quantum gate

$$\mathbf{T}(r) = r_1\mathbf{U}_1 + r_2\mathbf{U}_2 + r_3\mathbf{U}_3 + r_4\mathbf{U}_4 + r_5\mathbf{U}_5 + r_6\mathbf{U}_6 + r_7\mathbf{U}_7, \tag{4.58}$$

$$\mid r_1\mid + \mid r_2\mid + \mid r_3\mid + \mid r_4\mid + \mid r_5\mid + \mid r_6\mid + \mid r_7\mid = 1, \quad r_i \in \mathbb{C}. \tag{4.59}$$

If the parameters r_i are given, we can find the corresponding higher duality $n = 7$ slits quantum gate (4.57) with the composition of three (ternary) dividers (4.50) and two (4-ary) combiners (4.51), which have the following probability distributions $\left\{\vec{P}(3)\right\}$ and $\left\{\vec{Q}(4)\right\}$, where, e.g.,

$$p_1 = \frac{1}{r_2} r_4 \frac{r_5}{r_7}, \quad p_2 = \frac{r_1}{r_2 r_3^2} r_4 r_5 \frac{r_6}{r_7} \sqrt[6]{\frac{1}{r_1^5} r_2^3 \frac{r_3^8}{r_4 r_5^3 r_6^2} r_7}, \quad p_3 = \frac{r_1^2}{r_2 r_3^3} r_5 r_6 \sqrt[3]{\frac{1}{r_1^5} r_2^3 \frac{r_3^8}{r_4 r_5^3 r_6^2} r_7}, \tag{4.60}$$

$$q_1 = \frac{r_1^3}{r_2 r_3^4} r_5 r_6 \sqrt{\frac{1}{r_1^5} r_2^3 \frac{r_3^8}{r_4 r_5^3 r_6^2} r_7}, \quad q_2 = \frac{r_1}{r_3^2} r_5 \sqrt[3]{\frac{1}{r_1^5} r_2^3 \frac{r_3^8}{r_4 r_5^3 r_6^2} r_7}, \tag{4.61}$$

$$q_3 = \sqrt[6]{\frac{1}{r_1^5} r_2^3 \frac{r_3^8}{r_4 r_5^3 r_6^2} r_7}, \quad q_4 = \frac{1}{r_1} r_3 \frac{r_4}{r_6}, \tag{4.62}$$

and there are five other more cumbersome solutions. We should also take (4.59) into account, which gives (4.52).

Let us consider the reverse convolution (with respect to (4.39)) by the divider and combiner of an operator in $\mathcal{H}$.

Definition 4.26. The duality reverse n-ary convolution of the operator $\mathbf{T}$: $\mathcal{H} \to \mathcal{H}$ by the divider $\mathbf{D}_{\mathrm{p}}(n)$ and combiner $\mathbf{C}_{\mathrm{q}}(n)$ is the composition

$$\mathbf{P}^{(n),\mathrm{q},\mathrm{p}}_{\text{rdual}}(\mathbf{T}) = \mathbf{D}_{\mathrm{p}}(n)\circ\mathbf{T}\circ\mathbf{C}_{\mathrm{q}}(n). \tag{4.63}$$

Proposition 4.27. The action of the duality reverse n-ary convolution with the probability distributions $\left\{\vec{Q}\right\}$ and $\left\{\vec{P}\right\}$ on the vector quantum state $\left|\vec{\Psi}_{=}(n)\right\rangle \in \mathcal{H}$ is

$$\mathbf{P}_{\mathrm{rdual}}^{(n),\mathrm{q},\mathrm{p}}(\mathbf{T})\left|\vec{\Psi}_{=}(n)\right\rangle = \left(\vec{P}\vec{Q}^{T}\right) \otimes_{\mathrm{K}} \hat{T}\left|\vec{\Psi}_{=}(n)\right\rangle, \tag{4.64}$$

where $\otimes_{\mathrm{K}}$ is the Kronecker product of the matrix $\vec{P}\vec{Q}^{T} \in M_{n\times n}(\mathbb{C})$ and $\hat{T}$ is the matrix of the operator $\mathbf{T}$ in its matrix representation.

Proof. It follows from the manifest form of the divider (4.22) and combiner (4.27) and their linearity.

Note that if the duality reverse n-ary convolution (4.63) could be identity (for $\mathbf{T} = \mathrm{id}$), then together with (4.33) the divider $\mathbf{D}_{\mathrm{p}}(n)$ and combiner $\mathbf{C}_{\mathrm{q}}(n)$ become an n-ary analog of the biproduct in category theory (Mac Lane 1971). However, the condition that is needed for this condition ($\vec{P}\vec{Q}^{T}$ is the identity matrix) is never satisfied for nonvanishing probability distributions.

4.3 Duality quantum mode

The duality computer can be simulated by the ordinary quantum computer in a special work mode, i.e., having an additional/auxiliary qubit (or qudit) (Wei *et al* 2016). The main idea of the duality quantum mode computer is to provide the one-to-one correspondence of the auxiliary qudit state with the unitary operations on the slits (Long 2011).

The total state of the k-qubits $|\psi\rangle_{\mathrm{init}} \in \mathcal{H}$ and one auxiliary qudit (n-dit) representing n-slits $|\varphi\rangle_{\mathrm{aux}} \in \mathcal{H}$ is the direct product $|\psi\rangle_{\mathrm{init}} \otimes |\varphi\rangle_{\mathrm{aux}}$. The divider operation is represented by the unitary operator $\mathbf{V}$ acting on qudit, while the combider operation corresponds to the unitary operator $\mathbf{W}$ acting on qudit $|\varphi\rangle_{\mathrm{aux}}$. Between $\mathbf{V}$ and $\mathbf{W}$ there are n controlled operations corresponding to $\mathbf{U}_1, \mathbf{U}_2 \cdots \mathbf{U}_n$ one-to-one related to the states of qudit $\mathbf{U}_i \longleftrightarrow |i-1\rangle$. The whole duality quantum mode process can be presented as four consequent steps.

(1) Action of $\mathbf{V}$ on $|0\rangle_{\mathrm{aux}}$. Prepare the initial quantum state, e.g., with $|\varphi\rangle_{\mathrm{aux}} = |0\rangle_{\mathrm{aux}} = |0\rangle$, as $|\psi\rangle_{\mathrm{init}} \otimes |0\rangle$. The divider operation for n slits corresponds to the acting of the unitary operator $\mathbf{V}$ of the qudit (n-dit) state $|0\rangle$ as

$$|0\rangle \overset{\mathbf{V}}{\mapsto} \mathbf{V}|0\rangle = \overbrace{\left(\sum_{i=1}^{n} |i-1\rangle\langle i-1|\right)}^{\mathrm{id}} \mathbf{V}|0\rangle = \sum_{i=1}^{n} V_{i,1}|i-1\rangle, \tag{4.65}$$

where $V_{i,1} = \langle i-1|\mathbf{V}|0\rangle \in \mathbb{C}$ is the convolution of operator $\mathbf{V}\colon \mathcal{H} \to \mathcal{H}$ between the states $|i-1\rangle$ and $|0\rangle$, i.e., its matrix element, which represents the divider n-ary structure $\left\{\vec{P}(n)\right\}$ (4.23) after the identification

$$p_i = V_{i,1} = \langle i - 1 \mid \mathbf{V} \mid 0 \rangle. \tag{4.66}$$

Obviously, by definition

$$\sum_{i=1}^{n} | V_{i,1} | = \sum_{i=1}^{n} | p_i |^2 = 1, \tag{4.67}$$

as it should be for probabilities. Thus, using (4.65) and (4.66) the final sub-state $| \psi \rangle_{\text{init}} \otimes | i - 1 \rangle$ corresponds to the ith slit sub-wave, and

$$| \psi \rangle_{\text{init}} \otimes | 0 \rangle \overset{\text{id} \otimes \mathbf{V}}{\mapsto} | \psi \rangle_{\text{init}} \otimes \sum_{i=1}^{n} p_i \, | i - 1 \rangle = \sum_{i=1}^{n} p_i \, | \psi \rangle_{\text{init}} \otimes | i - 1 \rangle. \tag{4.68}$$

(2) Action of $\mathbf{U}_i$ on $| \psi \rangle_{\text{init}}$. The auxiliary controlled operation means that the action of the unitary operator $\mathbf{U}_i$ on $| \psi \rangle_{\text{init}}$ will be applied to the ith slit only, i.e., to the ith summand inside the last term (4.68). This is the reason why the same (for each i) initial state $| \psi \rangle_{\text{init}}$ was inserted into the sum. Therefore,

$$\sum_{i=1}^{n} p_i \, | \psi \rangle_{\text{init}} \otimes | i - 1 \rangle \overset{\mathbf{U}_i \otimes \text{id}}{\mapsto} \sum_{i=1}^{n} p_i (\mathbf{U}_i \, | \psi \rangle_{\text{init}}) \otimes | i - 1 \rangle. \tag{4.69}$$

(3) Action of $\mathbf{W}$ on the state $| i - 1 \rangle$. By analogy with (4.65) for each ith slit we have

$$| i - 1 \rangle \overset{\mathbf{W}}{\mapsto} \mathbf{W} \, | i - 1 \rangle = \overbrace{\left(\sum_{j=1}^{n} | j - 1 \rangle \langle j - 1 | \right)}^{\text{id}} \mathbf{W} \, | i - 1 \rangle = \sum_{j=1}^{n} W_{j,i} \, | j - 1 \rangle, \tag{4.70}$$

where $W_{j,i} = \langle j - 1 \mid \mathbf{W} \mid i - 1 \rangle \in \mathbb{C}$ is the convolution of operator (being a representative of the combiner) $\mathbf{W}\colon \mathcal{H} \to \mathcal{H}$ between the states $| j - 1 \rangle$ and $| i - 1 \rangle$, i.e., its matrix element, which represents (for fixed ith slit) the probabilities

$$q_j^{(i)} = W_{j,i} = \langle j - 1 \mid \mathbf{W} \mid i - 1 \rangle \in \mathbb{C}. \tag{4.71}$$

Now the normalization condition is

$$\sum_{i=1}^{n} | W_{j,i} | = \sum_{i=1}^{n} | q_j^{(i)} |^2 = 1, \quad \forall j = 1, \ldots, n, \tag{4.72}$$

as it should be by definition of probability. Then, using (4.69) and (4.70), the final quantum state will take the form

$$
\begin{aligned}
|\psi\rangle_{\text{init}} \otimes |0\rangle \mapsto & \sum_{i=1}^{n} p_i \mathbf{U}_i |\psi\rangle_{\text{init}} \otimes \sum_{j=1}^{n} q_j^{(i)} |j-1\rangle \\
= & \sum_{j=1}^{n}\sum_{i=1}^{n} p_i q_j^{(i)} \mathbf{U}_i(|\psi\rangle_{\text{init}} \otimes |j-1\rangle) = \sum_{j=1}^{n} \mathbf{T}_{\text{qdual}}^{(j)} |\psi\rangle_{\text{init}} \otimes |j-1\rangle,
\end{aligned}
\tag{4.73}
$$

where

$$
\mathbf{T}_{\text{qdual}}^{(j)} = \sum_{i=1}^{n} q_j^{(i)} p_i \mathbf{U}_i \tag{4.74}
$$

represents the duality gate in ordinary quantum computer.

(4) Complete measurement. After the previous three steps the auxiliary qubit arrives into the superposition state; therefore, the ith detector is placed at ith slit, when the qudit wave function is in the final state $|i-1\rangle$.

Consider the properties of the duality gate operator $\mathbf{T}_{\text{qdual}}^{(j)}$ (4.74). The condition

$$
\left| \sum_{i=1}^{n} q_j^{(i)} p_i \right| \leqslant 1, \quad \forall j = 1,\ldots,n, \tag{4.75}
$$

leads to the allowable duality gates (Long *et al* 2009). Because of the Cauchy–Bunyakovsky–Schwarz inequality and unitarity of the operators **V** and **W**, together with (4.67) and (4.72), we have

$$
\sum_{i=1}^{n} \left| q_j^{(i)} p_i \right| \leqslant 1, \quad \forall j = 1,\ldots,n. \tag{4.76}
$$

The duality gate (4.74) with the condition (4.76) is called the restricted allowable generalized quantum gate (Long *et al* 2009, Cao *et al* 2010).

Now we present the higher duality computation on the ordinary quantum computer. To model the polyadic power of the divider (4.44), we introduce the higher analog of the unitary operator action on the qudit state (4.65) and call it the duality power denoted by $(\ell_{\circledast})$. In this case the number of slits n does not coincide with the number of vectors of qudit n_d, and they are related by the formula of polyadic power (4.54) as follows

$$
n = \ell_{\circledast}(n_d - 1) + 1. \tag{4.77}
$$

We propose the following definition of the duality power, which is consistent with the polyadoc power (4.44). For instance, in case of quadratic duality power, instead of (4.65), we have for the n slits and n_d-dit the general formula

$$
|0\rangle \overset{\mathbf{V}^{\circledast 2}}{\mapsto} \mathbf{V}^{\circledast 2}|0\rangle = \sum_{i=1}^{n_d}\left[\delta_{i,1}V_{i,1}\left(\sum_{j=1}^{n_d} V_{j,1}|j-1\rangle\right) + (1-\delta_{i,1})V_{i,1}|i-1\rangle\right], \tag{4.78}
$$

$$n = 2n_d - 1, \tag{4.79}$$

where $\delta_{i,j}$ is the ordinary delta function, and $V_{i,1}$ is defined in (4.66). For the ternary duality power case $\ell_\circledast = 3$, to avoid cumbersome formulas, we give the concrete example $n_d = 3$, $n = 7$ from which the general pattern is clearly seen (cf (4.50))

$$|0\rangle \overset{\mathbf{V}^{\circledast 3}}{\mapsto} \mathbf{V}^{\circledast 3}|0\rangle = p_1\big[p_1\big(p_1|0\rangle + p_2|1\rangle + p_3|2\rangle\big) + p_2|1\rangle + p_3|2\rangle\big] + p_2|1\rangle + p_3|2\rangle, \tag{4.80}$$

where $p_i = V_{i,1}$ and $V_{i,j} = \langle i-1|\mathbf{V}|j-1\rangle$ is the matrix element of the unitary operator $\mathbf{V}$, $i, j = 1,\ldots,n_d$.

The other steps *(2)–(4)* in the duality quantum mode computation can be taken to be the same. This will lead to the higher nonlinear (in terms of the matrix elements of the operators $\mathbf{V}$ and $\mathbf{W}$) duality version of computation with the generalized nonunitary quantum gates of the form (4.74).

References

Abe E 1980 *Hopf Algebras* (Cambridge: Cambridge University Press)

Belousov V D 1972 *n-Ary Quasigroups* (Kishinev: Shtintsa)

Cao H, Chen Z, Guo Z and Ren F 2012 Complex duality quantum computers acting on pure and mixed states *Sci. China Phys. Mech. Astron.* **55** 2452–62

Cao H, Li L, Chen Z, Zhang Y and Guo Z 2010 Restricted allowable generalized quantum gates *Chin. Sci. Bull.* **55** 2122–5

Cao H-X, Long G-L, Guo Z-H and Chen Z-L 2013 Mathematical theory of generalized duality quantum computers acting on vector-states *Int. J. Theor. Phys.* **52** 1751–67

Chen L, Cao H-X and Meng H-X 2015 Generalized duality quantum computers acting on mixed states *Quantum Inf. Process.* **14** 4351–60

Duplij S 2022 *Polyadic Algebraic Structures* (Bristol: IOP Publishing)

Gudder S 2007 Mathematical theory of duality quantum computers *Quantum Inf. Process.* **6** 37–48

Gudder S 2008 Duality quantum computers and quantum operations *Int. J. Theor. Phys.* **47** 268–79

Lai C-Y, Brun T A and Wilde M M 2013 Duality in entanglement-assisted quantum error correction *IEEE Trans. Inf. Theory* **59** 4020–4

Liu Y 2013 Deleting a marked state in quantum database in a duality computing mode *Chin. Sci. Bull.* **58** 2927–31

Löhmus J, Paal E and Sorgsepp L 1994 *Nonassociative Algebras in Physics* (Palm Harbor: Hadronic Press)

Long G and Liu Y 2008 Duality quantum computing *Frontiers Comput. Sci. China* **2** 167–78

Long G-L 2006a General quantum interference principle and duality computer *Commun. Theor. Phys.* **45** 825–44

Long G L 2006b Mathematical theory of the duality computer in the density matrix formalism *Quantum Information Processing* **6** 2007

Long G L 2011 Duality quantum computing and duality quantum information processing *Int. J. Theor. Phys.* **50** 1305–18

Long G L and Liu Y 2007 Duality and recycling computing in quantum computers *Commun. Theor. Phys.* **50** 1303–6

Long G-L, Liu Y and Wang C 2009 Allowable generalized quantum gates *Commun. Theor. Phys.* **51** 65–7

Mac Lane S 1971 *Categories for the Working Mathematician* (Berlin: Springer)

Ohmori K 2008 Development of ultrahigh-precision coherent control and its applications *Proc. of the Japan Academy. Series B, Physical and Biological Sciences* **84** 167–75

Radford D E 2012 *Hopf Algebras* (Hackensack: World Scientific)

Sokhatsky F M 1997 On the associativity of multiplace operations *Quasigroups Relat. Syst.* **4** 51–66

Sweedler M E 1969 *Hopf Algebras* (New York: Benjamin)

Thurston H A 1949 Partly associative operations *J. London Math. Soc.* **24** 260–71

Wang W-Y, Shang B, Wang C and Long G L 2006 Prime factorization in the duality computer *Commun. Theor. Phys.* **47** 471–3

Wei S, Xin T and Long G 2017 Universal quantum channel simulation in duality quantum computing *IEEE Globecom Workshops* (Singapore: IEEE) pp 1–7

Wei S-J and Long G-L 2016 Duality quantum computer and the efficient quantum simulations *Quantum Inf. Process.* **15** 1189–212

Wei S-J, Ruan D and Long G-L 2016 Duality quantum algorithm efficiently simulates open quantum systems *Sci Rep.* **6** 30727

Zhang Q, Chen Z, Yuan F and Wang W 2018 A note of coherence for duality quantum computers acting on pure states *Int. J. Theor. Phys.* **57** 3795–807

Zheng C 2018 Duality quantum simulation of a general parity-time-symmetric two-level system *Europhys. Lett.* **123** 40002

Zheng C 2021 universal quantum simulation of single-qubit nonunitary operators using duality quantum algorithm *Sci. Rep.* **11** 1–14

Chapter 5

Measurement-based quantum computing

The measurement-based quantum computation model (Raussendorf and Briegel 2001, Raussendorf *et al* 2003) is a counterpart of the standard circuit model grounded on the unitary evolution (Deutsch 1989, Deutsch and Jozsa 1992). In the latter, the measurement is provided at the end of the whole computation to get the classical output, while in the measurement-based computation the principal operation is the measurement itself. Informally, the computation starts with several entangled qubits and measurements act on each qubit separately. The result is then exploited for next measurements. To avoid measurement indeterminacy, local unitary operations (named corrections) are implemented, which give the one-way computation (Raussendorf and Briegel 2001).

In general, the computation or the measurement pattern (program) consists of input and output sets of qubits connected with the sequence of basic commands. The patterns are then merged using tensor products and compositions (Danos *et al* 2007). The one-qubit measurement-based basic commands are:

(1) *Preparation* N_i of qubit i in the state $|+\rangle_i = (|\,0\rangle + |\,1\rangle)/\sqrt{2}$ (set N);
(2) *Entanglement* $E_{ij} = \wedge Z_{ij}$ (controlled-Z) of two qubits i and j (set E);
(3) *Measurement* $M_i(\alpha)$ of qubit i defined by projections on $|\pm_\alpha\rangle_i = (|\,0\rangle \pm e^{i\alpha}\,|\,1\rangle)/\sqrt{2} = P(\alpha)\,|\pm\rangle$, where $P(\alpha) = \mathrm{diag}(1, e^{i\alpha})$ is the phase operator, $0 \leqslant \alpha \leqslant 2\pi$ is the angle of measurement (set M);
(4) *Corrections* being the Pauli operators X_i and Z_i (set C).

The result of the measurement provided at qubit i is presented by *outcomes* $s_i = 0, 1 \in \mathbb{Z}_2$, where the convention is $s_i = 0$, if the initial qubit $|+\rangle_i$ after the measurement becomes $|+_\alpha\rangle_i$, while $s_i = 1$, if $|+\rangle_i$ collapses into $|-_\alpha\rangle_i$. After the measurement, the set of initial qubits $\{i\}$ produces the sum (in $\mathbb{Z}_2$) of individual outcomes $\sum s_i = s$, which is named *signal*, while the set $\{i\}$ is called the *domain* of the signal.

doi:10.1088/978-0-7503-5281-9ch5

The main idea of this method is some additional functional dependence of the corrections $X_i \to X_i^{(s)}$, $Z_i \to Z_i^{(s)}$ and measurements $M_i(\alpha) \to M_i^{(s,t)}(\alpha)$ from signals s, t. Because the signals are in $\mathbb{Z}_2$, the dependences from them become simple discrete functions

$$X_i^{(s)} = \begin{cases} I, & s = 0, \\ X_i, & s = 1 \end{cases}, \qquad Z_i^{(s)} = \begin{cases} I, & s = 0, \\ Z_i, & s = 1 \end{cases}, \tag{5.1}$$

$$M_i^{(s,t)}(\alpha) = M_i((-1)^s\alpha + \pi t) = \begin{cases} M_i(\alpha), & s = 0,\ t = 0, \\ M_i(-\alpha), & s = 1,\ t = 0, \\ M_i(\alpha + \pi), & s = 0,\ t = 1, \\ M_i(-\alpha + \pi), & s = 1,\ t = 1 \end{cases}. \tag{5.2}$$

The signal modification of measurements (5.2) can be expressed by the X- and Z-actions $\mathbf{A}_X$ and $\mathbf{A}_Z$ of conjugation with the Pauli matrices defined by (no summation)

$$\mathbf{A}_X \circ M_i(\alpha) \equiv X_i M_i(\alpha) X_i = M_i(-\alpha) = M_i^{(1,\,0)}(\alpha), \tag{5.3}$$

$$\mathbf{A}_Z \circ M_i(\alpha) \equiv Z_i M_i(\alpha) Z_i = M_i(\alpha + \pi) = M_i^{(0,\,1)}(\alpha). \tag{5.4}$$

The actions (5.3) and (5.4) commute because the addition of angles α is mod 2π. Since the measurements are destructive, the actions (5.3) and (5.4) can be simplified as follows

$$M_i(\alpha) X_i = M_i(-\alpha), \tag{5.5}$$

$$M_i(\alpha) Z_i = M_i(\alpha - \pi). \tag{5.6}$$

The signal domains of dependent commands give the set of such measurements that should be made before determination of the actual command value.

In general, the measurement pattern $\mathcal{P}$ is defined as

(1) Three sets:

  a. The computation space V of qubits and the associated quantum state space $\mathcal{H}$, which is $\otimes_{i \in \mathsf{V}} \mathbb{C}^2$.

  b. The pattern inputs $\mathsf{In} \in \mathsf{V}$ and outputs $\mathsf{Out} \in \mathsf{V}$ sets, together with their associated quantum state spaces $\mathcal{H}_{\mathsf{In}}$ and $\mathcal{H}_{\mathsf{Out}}$, correspondingly.

(2) Two injective maps $\mathbf{i}$: $\mathsf{In} \to \mathsf{V}$ and $\mathbf{o}$: $\mathsf{Out} \in \mathsf{V}$.

(3) The finite sequence of n commands $A_1, A_2, \ldots, A_{n-1}, A_n$ which act from the left to the right as $A_n A_{n-1} \cdots A_2 A_1$ on the pattern inputs from the set In.

In this notation, the measurement pattern becomes the map $\mathbf{P}$: $\mathsf{In} \to \mathsf{Out}$ and the pattern type is denoted by (V, In, Out). To simplify the notation, the sets of states

i(In) and **o**(Out) are denoted by the same letters In and Out, correspondingly. Providing a consequent pattern computation needs four conditions of definiteness

- **Def0** If outcome state is not measured, then no commands depend on it.
  Otherwise, one tries to apply a command depending on an outcome that is not known.
- **Def1** If a qubit is measured, then no commands act on it.
  If not, then one tries to execute a command on an already measured, and therefore changed, qubit.
- **Def2** If a qubit is not the input one but not prepared, then no commands act on it.
  Otherwise, one tries to apply a command to a not existing qubit.
- **Def3** If a qubit is not output, then it can be measured.
  In other words, since measurement consumes the qubits, this statement makes sure that the final state is in the output state.

If all of the statements are satisfied, then the conjunction **Def** = **Def0** ∧**Def1** ∧ **Def2** ∧ **Def3** will be used. It is important that a given pattern should satisfy **Def**, in general. The case when one exploits neither input not output qubits in a pattern corresponds to the auxiliary qubits, which considerably enlarges the space computation complexity. To avoid this, one should use as small number of the auxiliary qubits as possible. Moreover, one assumes that the inputs In and outputs Out can intersect, which can lead to simple unitaries implementations (Danos *et al* 2005).

The combination of patterns can be provided in two ways:

Composition If for two patterns defined by (V_1, In_1, Out_1) and (V_2, In_2, Out_2) we have $V_1 \cap V_2 = Out_1 = In_2$, then the composition $\mathscr{P} = \mathscr{P}_2 \circ \mathscr{P}_1$ can be given by (V, In, Out), where

$$V = V_1 \cup V_2, \quad In = In_1, \quad Out = Out_2, \tag{5.7}$$

and the commands are concatenated consequently.

Tensor product If the sets do not intersect $V_1 \cap V_2 = \varnothing$, then we can construct the tensor product pattern $\mathscr{P} = \mathscr{P}_2 \otimes \mathscr{P}_1$ which is defined by

$$V = V_1 \cup V_2, \quad In = In_1 \cup In_2, \quad Out = Out_1 \cup Out_2. \tag{5.8}$$

Since the sets V_1 and V_2 are disjoint, the commands from different patterns commute and are applied for qubits from different sets independently.

If V is initially not measured qubits (and then still active), then we denote the set V* as the measured qubits, which become classical bits. Therefore, the computation state space is

$$S = \Sigma_{VV*}\mathcal{H}_V \times \mathbb{Z}_2^{V*}, \tag{5.9}$$

where $\mathbb{Z}_2^{\mathsf{V}*}$ is outcome space of bits. We denote $\mathbf{\Gamma}$: $\mathsf{V}^* \to \mathbb{Z}_2$, and then the space (5.9) becomes the set of the quadruples

$$\mathsf{S} = \{(\mathsf{V}, \mathsf{V}^*, q, \mathbf{\Gamma})\} \tag{5.10}$$

(for short notation, the set of pairs $\mathsf{S} = \{(q, \mathbf{\Gamma})\}$), where $q \in \mathcal{H}_\mathsf{V}$ is a quantum state. The value of signals given by $\mathbf{\Gamma}$ are defined as $s_\Gamma = \sum_{i \in \mathsf{In}} \mathbf{\Gamma}(i)$ with the sum in $\mathbb{Z}_2$. If the outcome is empty, then the notation $\mathbb{Z}_2^{\varnothing}$ will be used. Some modification of the outcome map can be defined as follows

$$\mathbf{\Gamma}[k/i](j) = \begin{cases} k, & i = j \\ \mathbf{\Gamma}(j), & i \neq j, \end{cases}, \quad i, j, k \in \mathbb{Z}_2. \tag{5.11}$$

which maps $\mathbb{Z}_2^{\mathsf{V}*\cup\{i\}} \to \mathbb{Z}_2$. We can then write the action of the commands on the computation space (5.9) as (suppressing V, V^*)

$$\begin{array}{ccccc} & & (q \otimes |+\rangle_i, \mathbf{\Gamma}) & & \\ & & \uparrow N_i & & \\ (Y_i^{(s_\Gamma)} q, \mathbf{\Gamma}) & \xleftarrow{Y_i^{(s_\Gamma)}} & (q, \mathbf{\Gamma}) & \xrightarrow{X_i^{(s_\Gamma)}} & (X_i^{(s_\Gamma)} q, \mathbf{\Gamma}) \\ & & \downarrow E_{ij} & & \\ & & (\wedge Z_{ij} q, \mathbf{\Gamma}) & & \end{array} \tag{5.12}$$

and the action of the measurements (5.2) on the quadruples (5.10) as

$$(\mathsf{V} \cup \{i\}, \mathsf{V}^*, q, \mathbf{\Gamma}) \xrightarrow{M_i^{(s,t)}(\alpha)} \begin{cases} (\mathsf{V}, \mathsf{V}^* \cup \{i\}, \langle +_{\alpha_\Gamma}|_i\, q, \mathbf{\Gamma}[0/i]), \\ (\mathsf{V}, \mathsf{V}^* \cup \{i\}, \langle -_{\alpha_\Gamma}|_i\, q, \mathbf{\Gamma}[1/i]), \end{cases}, \tag{5.13}$$

using the modified α from (5.2) in the form $\alpha_\Gamma = (-1)^{s_\Gamma}\alpha + t_\Gamma \pi$.

For instance, the structure of the patterns with Pauli corrections X, Y in terms of the above commands can be formally (with suppressing all indices) written as (Mhalla *et al* 2022) $(\prod XYM)EN$.

In general, the execution of a pattern can be presented in the diagrammatic form (schematically)

$$\mathcal{H}_{\mathsf{In}} \longrightarrow \mathcal{H}_{\mathsf{In}} \times \mathbb{Z}_2^{\varnothing} \xrightarrow{\text{preparation}} \mathcal{H}_\mathsf{V} \times \mathbb{Z}_2^{\varnothing} \xrightarrow{A_1, A_2, \ldots, A_{n-1}, A_n} \mathcal{H}_{\mathsf{Out}} \times \mathbb{Z}_2^{\mathsf{V}*\mathsf{Out}} \longrightarrow \mathcal{H}_{\mathsf{Out}}. \tag{5.14}$$

The Brach map denoted by $\mathbf{B}_\mathbf{s}$ can be formally written as

$$\mathbf{B}_\mathbf{s} = \mathbf{C}_\mathbf{s} \circ \mathbf{M}_\mathbf{s} \circ \mathbf{U}, \tag{5.15}$$

where $\mathbf{U}$: $\mathcal{H}_{\mathsf{In}} \hookrightarrow \mathcal{H}_\mathsf{V}$ is a unitary embedding which is branch independent, $\mathbf{M}_\mathbf{s}$: $\mathcal{H}_\mathsf{V} \to \mathcal{H}_{\mathsf{Out}}$ is a projection being collection of measurements along the branch, and $\mathbf{C}_\mathbf{s}$ is a map corresponding to corrections on the output. Because all of the above maps are unitary, the resulting branch map is also unitary

$$\sum_{\mathbf{s}} \mathbf{B}_{\mathbf{s}}^{\dagger} \mathbf{B}_{\mathbf{s}} = \mathbf{I}, \tag{5.16}$$

where $\mathbf{I}$ is the identity matrix, and $\mathbf{s} \in \mathbb{Z}_2^n$. It follows from (5.16), that each pattern is presented by a positive map that preserves the trace (Danos *et al* 2007). A pattern is called deterministic if it is a positive trace preserving map and sends pure states to pure states.

References

Bauer M, Bernard D and Tilloy A 2015 Computing the rates of measurement-induced quantum jumps *J. Phys. A: Math. Theor.* **48** 15 Id/No 25ft02

Benjamin S C, Lovett B W and Smith J M 2009 Prospects for measurement-based quantum computing with solid state spins *Laser Photonics Rev.* **3** 556–74

Booth R I, Kissinger A, Markham D, Meignant C and Perdrix S 2023 Outcome determinism in measurement-based quantum computation with qudits *J. Phys. A: Math. Theor.* **56** 38 Id/No 115 303

Bravyi S and Raussendorf R 2007 Measurement-based quantum computation with the toric code states *Phys. Rev.* A **76** 022304

Browne D E, Kashefi E, Mhalla M and Perdrix S 2007 Generalized flow and determinism in measurement-based quantum computation *New J. Phys.* **9** 250

Danos V, D'Hondt E, Kashefi E and Panangaden P 2005 Distributed measurement-based quantum computation ed P Selinger *Proceedings of the 3rd International Workshop on Quantum Programming Languages vol 170 of Electronic Notes in Theoretical Computer Science* (Chicago: Elsevier) pp 73–94

Danos V, Kashefi E and Panangaden P 2005 Parsimonious and robust realizations of unitary maps in the one-way model *Phys. Rev.* A **72** 064301

Danos V, Kashefi E and Panangaden P 2007 The measurement calculus *J. ACM* **54** 45 8–es

Deutsch D 1989 Quantum computational networks *Proc. R. Soc. Lond. A: Math. Phys. Eng. Sci.* **425** 73–90

Deutsch D and Jozsa R 1992 Rapid solution of problems by quantum computation *Proc. R. Soc. Lond. A: Math. Phys. Eng. Sci.* **439** 553–8

Else D V, Bartlett S D and Doherty A C 2012 Symmetry protection of measurement-based quantum computation in ground states *New J. Phys.* **14** 113016

Else D V, Schwarz I, Bartlett S D and Doherty A C 2012 Symmetry-protected phases for measurement-based quantum computation *Phys. Rev. Lett.* **108** 240505

Ezziane Z 2010 Quantum computing measurement and intelligence *Int. J. Quantum Chem.* **110** 981–92

Ferguson R R, Dellantonio L, Al Balushi A, Jansen K, Dür W and Muschik C A 2021 Measurement-based variational quantum eigensolver *Phys. Rev. Lett.* **126** 220501

Frembs M, Roberts S, Campbell E T and Bartlett S D 2023 Hierarchies of resources for measurement-based quantum computation *New J. Phys.* **25** Paper No. 013 002, 24

Fujii K and Hayashi M 2017 Verifiable fault tolerance in measurement-based quantum computation *Phys. Rev.* A **96** 030301

Gross D and Eisert J 2007 Novel schemes for measurement-based quantum computation *Phys. Rev. Lett.* **98** 220503

Gross D, Eisert J, Schuch N and Perez-Garcia D 2007 Measurement-based quantum computation beyond the one-way model *Phys. Rev.* A **76** 052315

Hamrit N and Perdrix S 2015 Reversibility in extended measurement-based quantum computation *Reversible Computation. 7th Int. Conf. RC 2015, Grenoble, France, July 16–17, 2015. Proceedings* (Cham: Springer) pp 129–38

Hoban M J, Wallman J J, Anwar H, Usher N, Raussendorf R and Browne D E 2014 Measurement-based classical computation *Phys. Rev. Lett.* **112** 140505

Jorrand P and Perdrix S 2004 Unifying quantum computation with projective measurements only and one-way quantum computation *Proc. SPIE, Quantum Informatics (QI'04), vol. 5833, 44 (2005)* https://doi.org/10.1117/12.620302

Kashefi E and Pappa A 2017 Multiparty delegated quantum computing *Cryptography* **1** 12

Khalid U, Ur Rehman J and Shin H 2020 Measurement-based quantum correlations for quantum information processing *Sci. Rep.* **10** 2443

Kwek L C, Wei Z and Zeng B 2012 Measurement-based quantum computing with valence-bond-solids *Int. J. Mod Phys* B **26** 40 Id/No 1 230 002

Lanyon B P, Jurcevic P, Zwerger M, Hempel C, Martinez E A, Dür W, Briegel H J, Blatt R and Roos C F 2013 Measurement-based quantum computation with trapped ions *Phys. Rev. Lett.* **111** 210501

Menicucci N C 2014 Fault-tolerant measurement-based quantum computing with continuous-variable cluster states *Phys. Rev. Lett.* **112** 120504

Mhalla M, Perdrix S and Sanselme L 2022 *Characterising Determinism in MBQCS Involving Pauli Measurements* (CNRS Grenoble: Universite Grenoble Alpes) p (arxiv:2207.09368)

Miller J and Miyake A 2016 Hierarchy of universal entanglement in 2D measurement-based quantum computation *npj Quantum Inf.* **2** 16036

Morimae T 2014a Acausal measurement-based quantum computing *Phys. Rev.* A **90** 010101

Morimae T 2014b Basics and applications of measurement-based quantum computing *International Symposium on Information Theory and its Applications, ISITA 2014, Melbourne, Australia, October 26–29, 2014 (Piscataway, NJ)* (IEEE) pp 327–30

Morimae T 2016 Finding resource states of measurement-based quantum computing is harder than quantum computing *Phys. Rev.* A **96** 052308

Morimae T and Fujii K 2012 Blind topological measurement-based quantum computation *Nat. Commun.* **3** 1036

Musanna F and Kumar S 2020 A novel three-party quantum secret sharing scheme based on Bell state sequential measurements with application in quantum image sharing *Quantum Inf. Process.* **19** 21 Id/No 348

Perdrix S 2007 Towards minimal resources of measurement-based quantum computation *New J. Phys.* **9** 206

Perdrix S and Jorrand P 2004 *Measurement-based quantum turing machines and questions of universalities* https://doi.org/10.48550/arXiv.quant-ph/0404146

Ping Y, Gauger E M and Benjamin S C 2012 Measurement-based quantum computing with a spin ensemble coupled to a stripline cavity *New J. Phys.* **14** 17 Id/No 013 030

Raussendorf R 2013 Contextuality in measurement-based quantum computation *Phys. Rev.* A **88** 022322

Raussendorf R and Briegel H J 2001 A one-way quantum computer *Phys. Rev. Lett.* **86** 5188–91

Raussendorf R, Browne D E and Briegel H J 2003 Measurement-based quantum computation on cluster states *Phys. Rev.* A **68** 022312

Seo S, Seong J and Bae J 2022 Correlations in noisy measurements *Open Syst. Inf. Dyn.* **29** 12 d/ No 2 250 009

Shi-Hao Z, Xiang-Dong Z and Lü-Zhou L 2021 Research progress of measurement-based quantum computation *Acta Phys. Sin.* **70** 210301

Smith A M, Alsing P M, Lott G E and Fanto M L 2015 Translating non-trivial algorithms from the circuit model to the measurement-based quantum computing model *J. Mod. Opt* **62** 1746–54

Stock R and James D F V 2008 A scalable, high-speed measurement-based quantum computer using trapped ions *Phys. Rev. Lett.* **102** 170501

Suzuki Y, Watabe S, Kawabata S and Masuda S 2023 Measurement-based preparation of stable coherent states of a kerr parametric oscillator *Sci. Rep.* **13** 1–10

Trisetyarso A and Van Meter R 2010 Circuit design for a measurement-based quantum carry-lookahead adder *Int. J. Quantum Inf.* **8** 843–67

Usher N and Browne D E 2017a Noise in one-dimensional measurement-based quantum computing *Quant. Comput. Inf.* **17** 1372–97

Usher N and Browne D E 2017b Noise in one-dimentsional measurement-based quatum computing Quantum Inform *Quantum Inform. Compu.* **17** 1372–97

van den Nest M, Dür W, Vidal G and Briegel H J 2007 Classical simulation versus universality in measurement-based quantum computation *Phys. Rev.* A **75** 012337

van den Nest M, Miyake A, Dür W and Briegel H J 2006 universal resources for measurement-based quantum computation *Phys. Rev. Lett.* **97** 150504

Wei T-C 2021 Measurement-based quantum computation *Oxford Research Encyclopedia of Physics* 2021 https://doi.org/10.1093/acrefore/9780190871994.013.31

Xu Y-Y 2013 Measurement-based interference in quantum computation *Commun. Theor. Phys.* **60** 289–95

Zwerger M, Briegel H J and Dür W 2014 Hybrid architecture for encoded measurement-based quantum computation *Sci. Rep.* **4** 5364

IOP Publishing

Innovative Quantum Computing

Steven Duplij and Raimund Vogl

Chapter 6

Quantum walks

Quantum walks are the quantum counterpart of the classical random walks and they play an important role in the modelling of many phenomena, e.g., information spreading in complex networks (Noh and Rieger 2004), optimal search strategies (Lv *et al* 2002), genetic sequence location (van den Engh *et al* 1992), and chemical reactions (Gillespie 1977). The term 'quantum walks' was introduced in Aharonov *et al* (1993), but the idea to incorporate quantum effects to stochastic calculus appeared in Iche and Nozieres (1978), while the coherence effects in evolution of Brownian quantum particle were first considered in Schwinger (1961). The quantum analogies of classical random walks in discrete time and space were investigated in Godoy and Fujita (1992). The quantum cellular automata were introduced in Grössing and Zeilinger (1988), which appeared to be equivalent to the construction of Aharonov *et al* (1993), and which can be considered as one particle sector of the former; for a review, see (Arrighi 2019) and more general (Venegas-Andraca 2012). The connections between correlated classical random walks and quantum walks were given in Konno (2009) using matrix methods.

There are two models of quantum walks:

(1) Discrete quantum walks consist of two systems, called a walker and a coin, and the evolution unitary operator acts on them in discrete time steps.
(2) Continuous quantum walks consists of one quantum system called a walker, which 'walks' without time restrictions, which is described by the evolution operator (Hamiltonian) and the Schrödinger equation (Childs *et al* 2002).

The general topology in both cases can be described by discrete graphs.

6.1 Discrete quantum walks

In the case of discrete quantum walks on a line, the total quantum state consists of quantum states of the walker and the coin, i.e., the total Hilbert state $\mathcal{H}_{\text{tot}}$ becomes the direct product

doi:10.1088/978-0-7503-5281-9ch6

$$\mathcal{H}_{\text{tot}} = \mathcal{H}_{\text{coin}} \otimes \mathcal{H}_{\text{walk}}. \tag{6.1}$$

The position of the walker is described by the vector from the computational basis of the walker Hilbert space $\left| \psi_{\text{walk}} \right\rangle \in \mathcal{H}_{\text{walk}}$, which is infinite-dimensional and countable, such that the walker state $\left| \psi_{\text{walk}} \right\rangle$ is the quantum superposition

$$| \psi_{\text{walk}} \rangle = \sum_{\ell \in \mathbb{Z}} w_\ell \, | \, \ell \rangle_{\text{w}}, \quad \sum_{\ell \in \mathbb{Z}} w_\ell^2 = 1, \quad w_\ell \in \mathbb{C}. \tag{6.2}$$

In distinction to the classical coin, which can be in two states, the quantum s-state coin can be not only in s canonical basis states $| \, \mathbf{0} \rangle_{\text{c}}$, $| \, \mathbf{1} \rangle_{\text{c}}, \ldots, | \, \mathbf{s-1} \rangle_{\text{c}}$, but also in their quantum superposition

$$| \, \psi_{\text{coin}} \rangle = \sum_{j=0}^{s-1} c_j \, | \, \mathbf{j} \rangle_{\text{c}}, \quad \sum_{j=0}^{s-1} c_j^2 = 1, \quad c_j \in \mathbb{C}. \tag{6.3}$$

Usually, to be closer to the classical case, one puts $s = 2$. The total state of the quantum walk is given by

$$| \, \Psi_{\text{tot}} \rangle = | \, \psi_{\text{coin}} \rangle \otimes | \, \psi_{\text{walk}} \rangle, \tag{6.4}$$

and the initial total state, if to take $\left| \psi_{\text{walk}} \right\rangle_{\text{initial}} = | \, 0 \rangle_{\text{w}}$, becomes

$$| \, \Psi_{\text{tot}} \rangle_{\text{initial}} = | \, \psi_{\text{coin}} \rangle_{\text{initial}} \otimes | \, 0 \rangle_{\text{w}}. \tag{6.5}$$

In general, the total state can be written as

$$| \, \Psi_{\text{tot}} \rangle = \sum_{\ell \in \mathbb{Z}} \left(\varphi_{0,\ell} \, | \, \mathbf{0} \rangle_{\text{c}} \otimes | \, \ell \rangle_{\text{w}} + \varphi_{1,\ell} \, | \, \mathbf{1} \rangle_{\text{c}} \otimes | \, \ell \rangle_{\text{w}} \right), \tag{6.6}$$

$$\sum_{\ell \in \mathbb{Z}} \left(\left| \varphi_{0,\ell} \right|^2 + \left| \varphi_{1,\ell} \right|^2 \right) = 1 \quad \varphi_{0,\ell}, \varphi_{1,\ell} \in \mathbb{C}. \tag{6.7}$$

It follows from (6.2)–(6.3) that

$$\varphi_{j,\ell} = c_j w_\ell, \quad \ell \in \mathbb{Z}, \quad j = 0, 1, \tag{6.8}$$

and so the normalization condition (6.7) reduces one parameter from the set of ones describing the total state (6.6).

By analogy with the classical random walk, we need one operator to move the walker on the line and one operator to play the same role as the coin toss. In contrast to the classic case, where such an operator is represented by a stochastic matrix, in the case of the quantum walk evolution there is no room for randomness before measurement and it is represented by a unitary matrix, which acts as an internal rotation in the internal state space. The goal of the coin operator is to render the coin state in a superposition, while the randomness is introduced by making a measurement on the system after both evolution operators have been applied to the total quantum system for many times.

Thus, the evolution of a quantum walk is driven by the special composite action of two unitary operators: (1) in the first, a shift operator **S** acts in a combined total position-coin space $\mathcal{H}_{\text{tot}}$; (2) in the second, the coin operator **C** acts in the coin space $\mathcal{H}_{\text{coin}}$. In this way, the total evolution is described by the unitary operator **U**, which is defined by the main formula of the coined quantum walk concept

$$\mathbf{U} = \mathbf{S}\circ(\mathbf{C} \otimes \mathbf{I}_{\text{w}}), \tag{6.9}$$

$$\mathbf{S}\colon \mathcal{H}_{\text{coin}} \otimes \mathcal{H}_{\text{walk}} \to \mathcal{H}_{\text{coin}} \otimes \mathcal{H}_{\text{walk}},\quad \mathbf{C}\colon \mathcal{H}_{\text{coin}} \to \mathcal{H}_{\text{coin}},\quad \mathbf{U}\colon \mathcal{H}_{\text{tot}} \to \mathcal{H}_{\text{tot}}, \tag{6.10}$$

where $\mathbf{I}_{\text{w}} \in \mathcal{H}_{\text{walk}}$ is the unity of the walker space $\mathcal{H}_{\text{walk}}$.

If we consider the two-state coin $s = 2$ (6.6), then the operator **S** should act on the total quantum state (6.4) by shifts that are dependent from the coin state

$$\mathbf{S}\circ\left(|\,\mathbf{0}\rangle_{\text{c}} \otimes |\,\ell\rangle_{\text{w}}\right) = |\,\mathbf{0}\rangle_{\text{c}} \otimes |\,\ell + 1\rangle_{\text{w}}, \tag{6.11}$$

$$\mathbf{S}\circ\left(|\,\mathbf{1}\rangle_{\text{c}} \otimes |\,\ell\rangle_{\text{w}}\right) = |\,\mathbf{1}\rangle_{\text{c}} \otimes |\,\ell - 1\rangle_{\text{w}}. \tag{6.12}$$

This can be written in the unified form

$$\mathbf{S}\circ\left(|\,\mathbf{j}\rangle_{\text{c}} \otimes |\,\ell\rangle_{\text{w}}\right) = |\,\mathbf{j}\rangle_{\text{c}} \otimes |\,\ell + (-1)^{j}\rangle_{\text{w}}, \tag{6.13}$$

i.e., the shift operator depends on the coin state $\mathbf{S} = \mathbf{S}_j$. Therefore, in the computational basis, **S** can be presented using two projections in $\mathcal{H}_{\text{c}}$ as (the outer product representation)

$$\mathbf{S} = |\,\mathbf{0}\rangle_{\text{c}}\langle\mathbf{0}\,|_{\text{c}} \otimes \sum_{\ell\in\mathbb{Z}} |\,\ell + 1\rangle_{\text{w}}\langle\ell\,|_{\text{w}} + |\,\mathbf{1}\rangle_{\text{c}}\langle\mathbf{1}\,|_{\text{c}} \otimes \sum_{\ell\in\mathbb{Z}} |\,\ell - 1\rangle_{\text{w}}\langle\ell\,|_{\text{w}}, \tag{6.14}$$

which satisfies the required shifting properties in the walker space (6.11)–(6.12).

The coin operator **C** is an arbitrary element of the unitary group $\mathcal{U}(s)$, and for the two-state coin $s = 2$, and it can be represented by the four real parameter 2×2 complex matrix $\hat{C}$ of the form

$$\hat{C} = \hat{C}_{\alpha,\beta,\gamma,\theta} = \begin{pmatrix} a & b \\ c & d \end{pmatrix} = e^{i\gamma}\begin{pmatrix} e^{i\alpha}\cos\theta & e^{i\beta}\sin\theta \\ -e^{-i\beta}\sin\theta & e^{-i\alpha}\cos\theta \end{pmatrix},\quad a, b, c, d \in \mathbb{C},\ \ \alpha, \beta, \gamma, \theta \in \mathbb{R}. \tag{6.15}$$

In most cases, for quantum walks with two-state coin the Hadamard operator is widely used

$$\mathbf{C}_{\text{H}} = \frac{1}{\sqrt{2}}\left(|\,\mathbf{0}\rangle_{\text{c}}\langle\mathbf{0}\,|_{\text{c}} + |\,\mathbf{0}\rangle_{\text{c}}\langle\mathbf{1}\,|_{\text{c}} + |\,\mathbf{1}\rangle_{\text{c}}\langle\mathbf{0}\,|_{\text{c}} - |\,\mathbf{1}\rangle_{\text{c}}\langle\mathbf{1}\,|_{\text{c}}\right), \tag{6.16}$$

or in the matrix representation (6.15)

$$\hat{C}_{\text{H}} = \hat{C}_{\alpha=\frac{\pi}{2},\beta=\frac{\pi}{2},\gamma=\frac{\pi}{2},\theta=\frac{\pi}{4}} = \frac{1}{\sqrt{2}}\begin{pmatrix} 1 & 1 \\ 1 & -1 \end{pmatrix}. \tag{6.17}$$

The evolution of the total state (6.4) during the discrete time (=t) quantum walk after t steps $| \Psi_{tot}(t)\rangle$ is given by the application of the unitary operator (6.9) t times in the following way

$$| \Psi_{tot}(t)\rangle = \mathbf{U}^t \, | \Psi_{tot}(0)\rangle, \tag{6.18}$$

where $| \Psi_{tot}(0)\rangle = | \Psi_{tot}\rangle_{initial}$ (6.5).

Example 6.1. Using (6.9) and (6.16), we can get the first three steps for the Hadamard quantum walk with the two-state coin as

$$| \Psi_{tot}(1)\rangle = \frac{1}{\sqrt{2}} \, | \mathbf{0}\rangle_c \otimes | 1\rangle_w + \frac{1}{\sqrt{2}} \, | \mathbf{1}\rangle_c \otimes | -1\rangle_w, \tag{6.19}$$

$$| \Psi_{tot}(2)\rangle = -\frac{1}{2} \, | \mathbf{1}\rangle_c \otimes | -2\rangle_w + \frac{1}{2}(| \mathbf{0}\rangle_c + | \mathbf{1}\rangle_c) \otimes | 0\rangle_w + \frac{1}{2} \, | \mathbf{0}\rangle_c \otimes | 2\rangle_w \tag{6.20}$$

$$= \frac{1}{2} \, | \mathbf{0}\rangle_c \otimes (| 0\rangle_w + | 2\rangle_w) + \frac{1}{2} \, | \mathbf{1}\rangle_c \otimes (| 0\rangle_w - | -2\rangle_w), \tag{6.21}$$

$$| \Psi_{tot}(3)\rangle = \frac{1}{2\sqrt{2}} \, | \mathbf{1}\rangle_c \otimes | -3\rangle_w - \frac{1}{2\sqrt{2}} \, | \mathbf{0}\rangle_c \otimes | -1\rangle_w + \frac{1}{2\sqrt{2}}(2 \, | \mathbf{0}\rangle_c + | \mathbf{1}\rangle_c) \otimes | 1\rangle_w + \frac{1}{2\sqrt{2}} \, | \mathbf{0}\rangle_c \otimes | 3\rangle_w \tag{6.22}$$

$$= \frac{1}{2\sqrt{2}} \, | \mathbf{0}\rangle_c \otimes (-| -1\rangle_w + 2 \, | 1\rangle_w + | 3\rangle_w) + \frac{1}{2\sqrt{2}} \, | \mathbf{1}\rangle_c \otimes (| 1\rangle_w + | -3\rangle_w). \tag{6.23}$$

If the final state at the time t is known $\Psi_{tot}(t)$, then the standard way to describe the quantum walk is the partial measurement of the walker state probabilities (see, e.g. Portugal 2013).

However, we now have the tensor product of two spaces (6.1). Therefore, to have the complete description of the quantum walk, we propose to also consider the partial measurement of the (s-) coin state probabilities.

Let the total state at the time t (6.18) have the general form (see (6.6)–(6.8))

$$| \Psi_{tot}(t)\rangle = \sum_{\ell \in \mathbb{Z}} \sum_{j=0}^{s-1} \varphi_{j,\ell}(t) \, | \mathbf{j}\rangle_c \otimes | \ell\rangle_w, \tag{6.24}$$

$$\sum_{\ell \in \mathbb{Z}} \left| \varphi_{j,\ell}(t) \right|^2 = 1 \quad \varphi_{j,\ell} \in \mathbb{C}. \tag{6.25}$$

We denote the doubly partial probability of the state $| \mathbf{j}\rangle_c \otimes | \ell\rangle_w$ at time t by

$$p_{j,\ell}(t) = \left| \varphi_{j,\ell}(t) \right|^2, \quad \sum_{\ell \in \mathbb{Z}} \sum_{j=0}^{s-1} p_{j,\ell}(t) = 1. \tag{6.26}$$

Now we propose to characterize the quantum walk by two partial probability distributions:

(1) The walker probability distribution

$$p_\ell^{\text{walk}}(t) = \sum_{j=0}^{s-1} \left| \varphi_{j,\ell}(t) \right|^2, \tag{6.27}$$

$$\sum_{\ell \in \mathbb{Z}} p_\ell^{\text{walk}}(t) = 1. \tag{6.28}$$

(2) The coin probability distribution

$$p_j^{\text{coin}}(t) = \sum_{\ell \in \mathbb{Z}} \left| \varphi_{j,\ell}(t) \right|^2, \tag{6.29}$$

$$\sum_{j=0}^{s-1} p_j(t) = 1. \tag{6.30}$$

In the standard approach (Portugal 2013), only the first (walker) distribution (6.27) is usually considered: the time is fixed by $t = t_0$, and the graph $\left\{\ell,\, p_\ell^{\text{walk}}(t_0)\right\}$ is plotted. Nevertheless, the coin probability distribution (6.29) gives additional information about the quantum walk. To observe the difference between (6.27) and (6.29) concretely, we continue *example* 6.1 in detail.

Example 6.2. (*Example* 6.1 continued) Here we compute the walker and coin probabilities (6.27) and (6.29) for three steps $t = 1, 2, 3$ of the Hadamard walk $\Psi_{\text{tot}}(t)$ in (6.19)–(6.23).The formulas (6.19), (6.20), and (6.22) are convenient to use for the walker probabilities, and the formulas (6.19), (6.21), and (6.23) can be used for the coin probabilities. We derive the walker probabilities $p_\ell^{\text{walk}}(t)$ from (6.19)

$$p_{\ell=1}^{\text{walk}}(t = 1) = p_{\ell=|1\rangle_{\text{w}}}^{\text{walk}}(t = 1) = \left(\frac{1}{\sqrt{2}}\right)^2 = \frac{1}{2}, \tag{6.31}$$

$$p_{\ell=-1}^{\text{walk}}(t = 1) = p_{\ell=|-1\rangle_{\text{w}}}^{\text{walk}}(t = 1) = \left(\frac{1}{\sqrt{2}}\right)^2 = \frac{1}{2}, \tag{6.32}$$

and from (6.20) we obtain the symmetric distribution

$$p_{\ell=-2}^{\text{walk}}(t = 2) = p_{\ell=|-2\rangle_{\text{w}}}^{\text{walk}}(t = 2) = \left(\frac{1}{2}\right)^2 = \frac{1}{4}, \tag{6.33}$$

$$p_{\ell=0}^{\text{walk}}(t=2) = p_{\ell=|0\rangle_\text{w}}^{\text{walk}}(t=2) = \left(\frac{1}{2}\right)^2 + \left(\frac{1}{2}\right)^2 = \frac{1}{2}, \tag{6.34}$$

$$p_{\ell=2}^{\text{walk}}(t=2) = p_{\ell=|2\rangle_\text{w}}^{\text{walk}}(t=2) = \left(\frac{1}{2}\right)^2 = \frac{1}{4}. \tag{6.35}$$

The probability distribution $p_\ell^{\text{walk}}(t)$ for the third step $t = 3$ is nonsymmetric (6.22)

$$p_{\ell=-3}^{\text{walk}}(t=3) = p_{\ell=|-3\rangle_\text{w}}^{\text{walk}}(t=3) = \left(\frac{1}{2\sqrt{2}}\right)^2 = \frac{1}{8}, \tag{6.36}$$

$$p_{\ell=-1}^{\text{walk}}(t=3) = p_{\ell=|-1\rangle_\text{w}}^{\text{walk}}(t=3) = \left(-\frac{1}{2\sqrt{2}}\right)^2 = \frac{1}{8}, \tag{6.37}$$

$$p_{\ell=1}^{\text{walk}}(t=3) = p_{\ell=|1\rangle_\text{w}}^{\text{walk}}(t=3) = \left(2\frac{1}{2\sqrt{2}}\right)^2 + \left(\frac{1}{2\sqrt{2}}\right)^2 = \frac{5}{8}, \tag{6.38}$$

$$p_{\ell=3}^{\text{walk}}(t=3) = p_{\ell=|3\rangle_\text{w}}^{\text{walk}}(t=3) = \left(\frac{1}{2\sqrt{2}}\right)^2 = \frac{1}{8}, \tag{6.39}$$

as well as for further steps (times) $t > 3$.

For the coin probabilities $p_\ell^{\text{coin}}(t)$ we have from (6.19)

$$p_{j=0}^{\text{coin}}(t=1) = p_{j=|\mathbf{0}\rangle_\text{c}}^{\text{coin}}(t=1) = \left(\frac{1}{\sqrt{2}}\right)^2 = \frac{1}{2}, \tag{6.40}$$

$$p_{j=1}^{\text{coin}}(t=1) = p_{j=|\mathbf{1}\rangle_\text{c}}^{\text{coin}}(t=1) = \left(\frac{1}{\sqrt{2}}\right)^2 = \frac{1}{2}, \tag{6.41}$$

and from (6.21) we have for the second step $t = 2$ the symmetric distribution

$$p_{j=0}^{\text{coin}}(t=2) = p_{j=|\mathbf{0}\rangle_\text{c}}^{\text{coin}}(t=2) = \left(\left(\frac{1}{2}\right)^2 + \left(\frac{1}{2}\right)^2\right) = \frac{1}{2}, \tag{6.42}$$

$$p_{j=1}^{\text{coin}}(t=2) = p_{j=|\mathbf{1}\rangle_\text{c}}^{\text{coin}}(t=2) = \left(\left(\frac{1}{2}\right)^2 + \left(-\frac{1}{2}\right)^2\right) = \frac{1}{2}, \tag{6.43}$$

The probability distribution $p_j^{\text{coin}}(t)$ for the third step $t = 3$ is also nonsymmetric as $p_\ell^{\text{walk}}(t = 3)$, so from (6.23) we get

$$p_{j=0}^{\text{coin}}(t=3) = p_{j=|\mathbf{0}\rangle_c}^{\text{coin}}(t=3) = \left(\left(-\frac{1}{2\sqrt{2}}\right)^2 + \left(2\frac{1}{2\sqrt{2}}\right)^2 + \left(\frac{1}{2\sqrt{2}}\right)^2\right) = \frac{3}{4}, \quad (6.44)$$

$$p_{j=1}^{\text{coin}}(t=3) = p_{j=|\mathbf{1}\rangle_c}^{\text{coin}}(t=3) = \left(\left(\frac{1}{2\sqrt{2}}\right)^2 + \left(2\frac{1}{2\sqrt{2}}\right)^2\right) = \frac{1}{4}, \quad (6.45)$$

and in the similar way for further steps (discrete times) $t > 3$.

As it should be, both the above walker and coin probability distributions are correctly normalized satisfying (6.28) and (6.30) at each discrete time t.

6.1.1 Polyander visualization of quantum walks

The coin probability distribution $p_j^{\text{coin}}(t)$ introducted in (6.29), from the first glance, can be also characterized at the fixed time $t = t_0$ by the graph $\left\{j, p_j^{\text{coin}}(t_0)\right\}$ as the walker probability distribution $p_\ell^{\text{walk}}(t_0)$. However, because the coin has a specific physical sense, we propose here another way of the quantum walk description, which originates from genome landscapes (Azbel' M Y 1973, 1995, Lobry 1996) and one-dimensional DNA walks (Cebrat and Dudek 1998) and trianders (Duplij and Duplij 2005).

Innovation 6.3. *We can consider the time evolution of the probability for the concrete quantum state when we provide the corresponding measurements in the coin or walker subspaces, i.e., we fix the states $\ell = \ell_0$ or $j = j_0$ and introduce the following time evolution graphs $\left\{t, p_{\ell=\ell_0}^{\text{walk}}(t)\right\}$ or $\left\{t, p_{j=j_0}^{\text{coin}}(t)\right\}$.*

Definition 6.4. The polyander visualization of a quantum walk is its description by the time evolution graphs $\left\{t, p_\ell^{\text{walk}}(t)\right\}$ or $\left\{t, p_j^{\text{coin}}(t)\right\}$. Each line of the graph describing the probability evolution of the fixed quantum state $\ell = \ell_0$ for $|\,\ell_0\rangle_w$ or $j = j_0$ for $|\,\mathbf{j}\rangle_c$ is called a leg of the polyander.

It is obvious that the walker polyander has a finitely increasing number of legs and corresponding quantum states, while the s-side coin polyander has exactly s legs.

For *example* 6.1, we obtain the following.

Example 6.5. (*Example* 6.1 continued) The walker polyander $p_\ell^{\text{walk}}(t)$ in the time range $1 \leqslant t \leqslant 3$ has seven legs (quantum states) $-3 \leqslant \ell \leqslant 3$, which have the following probability evolutions

$\vert\ell\rangle$-leg\time t	1	2	3
$\vert-3\rangle_w$	0	0	$\frac{1}{8}$
$\vert-2\rangle_w$	0	$\frac{1}{4}$	0
$\vert-1\rangle_w$	$\frac{1}{2}$	0	$\frac{1}{8}$
$\vert0\rangle_w$	0	$\frac{1}{2}$	0
$\vert1\rangle_w$	$\frac{1}{2}$	0	$\frac{5}{8}$
$\vert2\rangle_w$	0	$\frac{1}{4}$	0
$\vert3\rangle_w$	0	0	$\frac{1}{8}$

(6.46)

The coin polyander $p_j^{\text{coin}}(t)$ in the time range $1 \leqslant t \leqslant 3$ has two legs (quantum states), $j = 0, 1$, which have the following probability evolutions

$\vert\mathbf{j}\rangle$-leg\time t	1	2	3
$\vert\mathbf{0}\rangle_c$	$\frac{1}{2}$	$\frac{1}{2}$	$\frac{3}{4}$
$\vert\mathbf{1}\rangle_c$	$\frac{1}{2}$	$\frac{1}{2}$	$\frac{1}{4}$

(6.47)

Each leg can be presented as a horizontal strip of the width 1 on which the points corresponding to the probabilities $0 \leqslant p(t) \leqslant 1$ at times $t = 1, 2, 3...$ are indicated. The probability behaviour of each quantum state can then be visually seen and mutually compared in the same time points.

For the coin polyander, it is important to consider the probability differences because of the following

Definition 6.6. The total quantum state is called trivial at the time $t = t_{\text{triv}}$ if all the s-side coin states have equal probabilities $p_j^{\text{coin}}(t_{\text{triv}}) = \frac{1}{s}$, $j = 0, 1,...,s - 1$, $s \geqslant 2$.

Definition 6.7. The quantum walk is called trivial if the s-side coin states are trivial at all times.

In the case of the standard coin $s = 2$, the triviality means that the measurements of both sides give the same probability at the $t = t_{\text{triv}}$. Therefore, to describe triviality in detail, we should introduce the differences and search for nonzero ones.

Definition 6.8. The bias s-side coin polyander has $(s - 1)$ legs, which are defined by

$$\Delta p_j^{\text{coin}}(t) = p_j^{\text{coin}}(t) - p_{j+1}^{\text{coin}}(t), \quad j = 0,...,s - 2. \tag{6.48}$$

Example 6.9. (*Example* 6.1 continued) The 2-side coin bias polyander in the time range $1 \leqslant t \leqslant 3$ has one leg which has the following probability evolution $\Delta p_0^{\text{coin}}(t) = \Delta p_{j=|0\rangle_c}^{\text{coin}}(t) - \Delta p_{j=|1\rangle_c}^{\text{coin}}(t)$ (see (6.47))

$j \backslash$time t	1	2	3
0	0	0	$\frac{1}{2}$

(6.49)

which can be nontrivial after the time $t = 3$ only.

In the higher times, the walker and coin polyanders, as well as the bias coin polyander, will have more complicated behavior, which in any case needs the manifest form of the total quantum state (6.18). In *examples* 6.5 and 6.9, we considered for clarity only the time range $1 \leqslant t \leqslant 3$ and the 2-side coin to show in detail how to compute probability polyanders for finite times. The physical sense of the bias polyander is in the following: its nonzero values show nontriviality evolution along the quantum walk.

Thus, polyanders allow us to further study the fine structure, and thoroughly characterize and visually present quantum walks from different viewpoints.

6.1.2 Methods of final states computation

The main goal of studying the quantum walks is to obtain the analytical expression for the final quantum state (6.18) in discrete finite times $t \in \mathbb{Z}$, and then calculate the dynamical and statistical properties of various probability distributions and characteristics.

The main computational methods to find the total quantum state (6.18) are

1. **The Schrödinger approach**. Starting from an arbitrary state of the quantum walk with a certain walker position, to provide the discrete time Fourier transform (Ambainis *et al* 2001) and obtain the closed form of total amplitudes.
2. **The combinatorial approach**. The amplitude at any discrete time is derived as a sum of amplitudes of all paths starting from the initial state and ending up in the final state. This can be treated as reminiscent of the standard path integral technique.

In Carteret *et al* (2005), it was shown that both Schrödinger and combinatorial approaches are equivalent. Among less known methods, we can mention the alternative description of quantum walks based on the scattering theory (Feldman and Hillery 2007) and the analytic formulation of probability densities and moments (Fuss *et al* 2007).

6.1.2.1 Fourier transform and analytic solutions

In general, the usage of the Fourier transform is the standard way to simplify computations by turning equations to algebraic ones. In its application to quantum works and analysing the evolution (6.18), there two peculiarities:

(1) The Fourier transform is applied to one subspace from the product (6.4), i.e., the walker one $\mathcal{H}_{\text{walk}}$.
(2) Sometimes it is simpler to turn from transforming functions to transform the computational basis of the walker subspace.

Following *(2)*, we transform the computational basis of the walker space $\mathcal{H}_{\text{walk}}$ as

$$\| \mathsf{k}\rangle\rangle_{\mathrm{w}} = \sum_{\ell \in \mathbb{Z}} e^{i\mathsf{k}\ell} \,|\, \ell\rangle_{\mathrm{w}}, \quad \ell \in \mathbb{Z}, \ |\, \ell\rangle_{\mathrm{w}}, \ \| \mathsf{k}\rangle\rangle_{\mathrm{w}} \in \mathcal{H}_{\text{walk}}, \tag{6.50}$$

where the Fourier transformed vectors $\| \mathsf{k}\rangle\rangle_{\mathrm{w}}$ are denoted by the double brackets and depend on the continuous real wave number $\mathsf{k} \in \mathbb{R}$, $-\pi \leqslant \mathsf{k} \leqslant \pi$. The inverse transformation is

$$|\, \ell\rangle_{\mathrm{w}} = \frac{1}{2\pi} \int_{-\pi}^{\pi} d\mathsf{k} e^{-i\mathsf{k}\ell} \, \| \mathsf{k}\rangle\rangle_{\mathrm{w}}. \tag{6.51}$$

Let us introduce the Fourier transformation of the amplitudes $\varphi_{j,\ell}(t)$ at time t from the decomposition (6.24) in the standard way by

$$\Phi_{j,\mathsf{k}}(t) = \sum_{\ell \in \mathbb{Z}} e^{-i\mathsf{k}\ell} \varphi_{j,\ell}(t), \quad -\pi \leqslant \mathsf{k} \leqslant \pi. \tag{6.52}$$

The inverse Fourier transform becomes

$$\varphi_{j,\ell}(t) = \frac{1}{2\pi} \int_{-\pi}^{\pi} d\mathsf{k} e^{i\mathsf{k}\ell} \Phi_{j,\mathsf{k}}(t). \tag{6.53}$$

Then, instead of the computational basis $|\, \mathbf{j}\rangle_{\mathrm{c}} \otimes |\, \ell\rangle_{\mathrm{w}}$ in (6.24), using (6.50) and (6.53) and cancelling exponents, we can present the total state in the Fourier basis $|\, \mathbf{j}\rangle_{\mathrm{c}} \otimes \| \mathsf{k}\rangle\rangle_{\mathrm{w}}$ as follows

$$|\, \Psi_{\text{tot}}(t)\rangle = \frac{1}{2\pi} \sum_{j=0}^{s-1} \int_{-\pi}^{\pi} \Phi_{j,\mathsf{k}}(t) \,|\, \mathbf{j}\rangle_{\mathrm{c}} \otimes \| \mathsf{k}\rangle\rangle_{\mathrm{w}}. \tag{6.54}$$

The action of the shift operator $\mathbf{S}$ on the Fourier basis can be derived from (6.13) and using (6.50), as follows

$$\begin{aligned}
\mathbf{S}\circ(|\, \mathbf{j}\rangle_{\mathrm{c}} \otimes \| \mathsf{k}\rangle\rangle_{\mathrm{w}}) &= \sum_{\ell \in \mathbb{Z}} e^{i\mathsf{k}\ell} \mathbf{S}\circ(|\, \mathbf{j}\rangle_{\mathrm{c}} \otimes |\, \ell\rangle_{\mathrm{w}}) = \sum_{\ell \in \mathbb{Z}} e^{i\mathsf{k}\ell} \mathbf{S}\circ(|\, \mathbf{j}\rangle_{\mathrm{c}} \otimes |\, \ell\rangle_{\mathrm{w}}) \\
&= \sum_{\ell \in \mathbb{Z}} e^{i\mathsf{k}\ell}\left(|\, \mathbf{j}\rangle_{\mathrm{c}} \otimes |\, \ell + (-1)^j\rangle_{\mathrm{w}}\right) = \sum_{\ell' \in \mathbb{Z}} e^{i\mathsf{k}(\ell' - (-1)^j)}(|\, \mathbf{j}\rangle_{\mathrm{c}} \otimes |\, \ell'\rangle_{\mathrm{w}}) \\
&= e^{-i\mathsf{k}(-1)^j} \sum_{\ell' \in \mathbb{Z}} e^{i\mathsf{k}\ell'}(|\, \mathbf{j}\rangle_{\mathrm{c}} \otimes |\, \ell'\rangle_{\mathrm{w}}) = e^{-i\mathsf{k}(-1)^j} \,|\, \mathbf{j}\rangle_{\mathrm{c}} \otimes \| \mathsf{k}\rangle\rangle_{\mathrm{w}},
\end{aligned} \tag{6.55}$$

where we used the substitution $\ell' = \ell + (-1)^j$ and the translation symmetry of the infinite sum.

In the case of the two-side coin $j = 0, 1$ and the Hadamard quantum walk (6.16)–(6.17), the action of operators can be expressed in the matrix form.

So we apply the total evolution operator **U** (6.9) in the matrix form to the Fourier basis $|\,\mathbf{j}\rangle_c \otimes ||\,\mathsf{k}\rangle\rangle_w$ using (6.17) to get

$$\hat{U}(|\,\mathbf{j}'\rangle_c \otimes ||\,\mathsf{k}\rangle\rangle_w) = \hat{S}\left(\left(\sum_{j=0}^{1} \hat{C}_{jj'}\,|\,\mathbf{j}\rangle_c\right) \otimes ||\,\mathsf{k}\rangle\rangle_w\right)$$
$$= \left(\sum_{j=0}^{1} e^{-i\mathsf{k}(-1)^j}\hat{C}_{jj'}\,|\,\mathbf{j}\rangle_c\right) \otimes ||\,\mathsf{k}\rangle\rangle_w = \sum_{j=0}^{1} \bar{\mathbf{C}}_{jj'}(\mathsf{k})\,|\,\mathbf{j}\rangle_c \otimes ||\,\mathsf{k}\rangle\rangle_w, \tag{6.56}$$

where

$$\bar{\mathbf{C}}(\mathsf{k}) = \begin{pmatrix} e^{-i\mathsf{k}} & 0 \\ 0 & e^{i\mathsf{k}} \end{pmatrix}\hat{C} = \frac{1}{\sqrt{2}}\begin{pmatrix} e^{-i\mathsf{k}} & e^{-i\mathsf{k}} \\ e^{i\mathsf{k}} & -\,e^{i\mathsf{k}} \end{pmatrix}. \tag{6.57}$$

It follows from (6.56) that diagonalization of $\bar{\mathbf{C}}(\mathsf{k})$ leads to the spectral decomposition of the total operator $\hat{U}$. Indeed, if $\lambda(\mathsf{k})$ is the eigenvalue of the matrix $\bar{\mathbf{C}}(\mathsf{k})$, then it is also the eigenvalue of $\hat{U}$, as is seen from (6.56). We denote the corresponding $\lambda(\mathsf{k})$ eigenvector by $\left|\left|\, \mathsf{v}_{\lambda(\mathsf{k})}\right\rangle\right\rangle_c$, such that

$$\hat{U}\circ\left(\left|\left|\, \mathsf{v}_{\lambda(\mathsf{k})}\right\rangle\right\rangle_c \otimes ||\,\mathsf{k}\rangle\rangle_w\right) = \left(\bar{\mathbf{C}}(\mathsf{k})\circ\left|\left|\, \mathsf{v}_{\lambda(\mathsf{k})}\right\rangle\right\rangle_c\right) \otimes ||\,\mathsf{k}\rangle\rangle_w = \lambda(\mathsf{k})\left|\left|\, \mathsf{v}_{\lambda(\mathsf{k})}\right\rangle\right\rangle_c \otimes ||\,\mathsf{k}\rangle\rangle_w. \tag{6.58}$$

The matrix $\bar{\mathbf{C}}(\mathsf{k})$ (6.57) has two eigenvalues

$$\lambda_1(\mathsf{k}) = e^{-i\alpha(\mathsf{k})}, \quad \lambda_2(\mathsf{k}) = -e^{i\alpha(\mathsf{k})}, \tag{6.59}$$

$$\alpha(\mathsf{k}) = \arcsin\left(\frac{1}{\sqrt{2}}\sin\mathsf{k}\right), \; -\frac{\pi}{2} \leqslant \alpha(\mathsf{k}) \leqslant \frac{\pi}{2}, \tag{6.60}$$

and two corresponding normalized eigenvectors

$$\left|\left|\, \mathsf{v}_{\lambda_{1,2}(\mathsf{k})}\right\rangle\right\rangle_c = \frac{1}{\sqrt{r_{1,2}}}\begin{pmatrix} e^{-i\mathsf{k}} \\ \pm\sqrt{2}\,e^{-i\alpha(\mathsf{k})} - e^{-i\mathsf{k}} \end{pmatrix}, \tag{6.61}$$

$$r_{1,2} = 2\left(1 + \cos^2\mathsf{k} \mp \cos\mathsf{k}\sqrt{1 + \cos^2\mathsf{k}}\right). \tag{6.62}$$

Thus, in the total evolution the operator can be written in terms of eigenvalues and eigenvectors of $\bar{\mathbf{C}}(\mathsf{k})$ (6.57)

$$\hat{U} = \frac{1}{2\pi}\int_{-\pi}^{\pi} d\mathsf{k}\left[\left(e^{-i\alpha(\mathsf{k})}\left|\left|\, \mathsf{v}_{\lambda_1(k)}\right\rangle\right\rangle_c\left\langle\left\langle \mathsf{v}_{\lambda_1(k)}\right|\right|_c - e^{i\alpha(\mathsf{k})}\left|\left|\, \mathsf{v}_{\lambda_2(k)}\right\rangle\right\rangle_c\left\langle\left\langle \mathsf{v}_{\lambda_2(k)}\right|\right|_c\right) \otimes ||\,\mathsf{k}\rangle\rangle_w\langle\langle \mathsf{k}\,||_w\right]. \tag{6.63}$$

Using orthogonality of the basis eigenvectors, the power of the evolution operator can be presented as

$$\hat{U}^t = \frac{1}{2\pi}\int_{-\pi}^{\pi} d\mathsf{k}\left[\left(e^{-i\alpha(\mathsf{k})t}\left\|\mathsf{v}_{\lambda_1(k)}\right\rangle\!\right\rangle_{\mathsf{c}}\left\langle\!\left\langle \mathsf{v}_{\lambda_1(k)}\right\|_{\mathsf{c}} + (-1)^t e^{i\alpha(\mathsf{k})t}\left\|\mathsf{v}_{\lambda_2(k)}\right\rangle\!\right\rangle_{\mathsf{c}}\left\langle\!\left\langle \mathsf{v}_{\lambda_2(k)}\right\|_{\mathsf{c}}\right) \otimes \| \mathsf{k}\rangle\!\rangle_{\mathsf{w}}\langle\!\langle \mathsf{k}\|_{\mathsf{w}}\right]. \tag{6.64}$$

Now we can use the main quantum evolution formula (6.18) to obtain the total quantum state at any time from an initial quantum state (6.5). For instance, if $|\,\Psi_{\text{tot}}\rangle_{initial} = |\,\mathbf{0}\rangle_{\text{c}} \otimes |\,0\rangle_{\text{w}}$, then using (6.54) and (6.61), we derive the Fourier transformed amplitudes

$$\begin{aligned}\Phi_{j=0,\mathsf{k}}(t) &= \frac{1}{2\sqrt{1+\cos^2\mathsf{k}}}\left[\left(\sqrt{1+\cos^2\mathsf{k}}+\cos\mathsf{k}\right)e^{-i\alpha(\mathsf{k})t} + \left(\sqrt{1+\cos^2\mathsf{k}}-\cos\mathsf{k}\right)e^{i(\pi+\alpha(\mathsf{k}))t}\right],\\ \Phi_{j=1,\mathsf{k}}(t) &= \frac{e^{i\mathsf{k}}}{2\sqrt{1+\cos^2\mathsf{k}}}\left(e^{-i\alpha(\mathsf{k})t} - e^{i(\pi+\alpha(\mathsf{k}))t}\right).\end{aligned} \tag{6.65}$$

Then by applying the reverse Fourier transform (6.53) and taking into account symmetries of integrand, we get the amplitudes in the computational basis at the arbitrary time t as

$$\varphi_{j=0,\ell}(t) = \begin{cases}\dfrac{1}{2\pi}\displaystyle\int_{-\pi}^{\pi} d\mathsf{k}e^{i(\mathsf{k}\ell-\alpha(\mathsf{k})t)}\left(\dfrac{\cos\mathsf{k}}{\sqrt{1+\cos^2\mathsf{k}}}+1\right), & t+\ell = \text{even},\\ 0, \quad t+\ell = \text{odd},\end{cases} \tag{6.66}$$

$$\varphi_{j=1,\ell}(t) = \begin{cases}\dfrac{1}{2\pi}\displaystyle\int_{-\pi}^{\pi} d\mathsf{k}e^{i(\mathsf{k}\ell-\alpha(\mathsf{k})t+\mathsf{k})}\dfrac{1}{\sqrt{1+\cos^2\mathsf{k}}}, & t+\ell = \text{even},\\ 0, \quad t+\ell = \text{odd}.\end{cases} \tag{6.67}$$

Finally, using the partial probability formulas (6.27) and (6.29), one can plot the time evolution graphs $\left\{t,\, p_{\ell=\ell_0}^{\text{walk}}(t)\right\}$ and $\left\{t,\, p_{j=j_0}^{\text{coin}}(t)\right\}$, i.e., to provide the polyander visualization (see section 6.1.1).

6.1.3 Generalizations of discrete-time quantum walks

There are plenty of generalizations of the above constructions. Nevertheless, the main procedures remain nearly the same.

- **Coin operator.** The most general form of the two-sided ($s = 2$) coin operator $\mathbf{C}$ is given by the complex matrix (6.15) from the unitary group $\mathscr{U}(2)$, i.e., other than the Hadamard matrix (6.17) can be considered, such as the Fourier coin (Portugal 2013).
- **Higher dimensions.** The main quantum walk equation (6.9) can be extended to a higher dimension of the s-sided coin when $\mathcal{H}_{\text{coin}}$ is $2s$-dimensional Hilbert space and $\mathcal{H}_{\text{walk}}$ is the Hilbert space corresponding to the direct product $\overbrace{\mathbb{Z}\otimes\cdots\otimes\mathbb{Z}}^{s}$. The common choice for an s-sided coin is the Grover operator

described by the corresponding $2s$-dimensional matrix $\hat{C}_{\text{Grover}}$ that was proposed in Moore and Russell (2002).

- **Anyonic quantum walks.** To include the braiding interaction, one includes the additional Hilbert space (fusion space) $\mathcal{H}_{\text{fusion}}$ where the generators of the braid group act. Then the total space becomes $\mathcal{H}_{\text{tot}} = \mathcal{H}_{\text{coin}} \otimes \mathcal{H}_{\text{fusion}} \otimes \mathcal{H}_{\text{walk}}$, and the time evolution contains the additional braid operator in some representation (Lehman *et al* 2011).

References

Abal G, Donangelo R, Forets M and Portugal R 2012 Spatial quantum search in a triangular network *Math. Struct. Comput. Sci.* **22** 521–31

Acasiete F, Agostini F P, Moqadam J K and Portugal R 2020 *Quantum Inf. Process.* **19** 426

Acevedo O L, Roland J and Cerf N J 2008 Exploring scalar quantum walks on cayley graphs *Quantum Inf. Comput.* **8** 68–81

Aharonov Y, Davidovich L and Zagury N 1993 Quantum random walks *Phys. Rev.* A **48** 1687–90

Ahlbrecht A, Vogts H, Werner A H and Werner R F 2011 Asymptotic evolution of quantum walks with random coin *J. Math. Phys.* **52** 36

Ahmad R, Sajjad U and Sajid M 2020 One-dimensional quantum walks with a position-dependent coin *Commun. Theor. Phys.* **11** Id/No 065 101

Alberti A, Alt W, Werner R and Meschede D 2014 Decoherence models for discrete-time quantum walks and their application to neutral atom experiments *New J. Phys.* **16** 123052

Allegrini P, Grigolini P and West B J 1996 A dynamical approach to DNA sequences *Phys. Lett.* A **211** 217–22

Alvir R, Dever S, Lovitz B, Myer J, Tamon C, Xu Y and Zhan H 2016 Perfect state transfer in Laplacian quantum walk *J. Algebraic Combin* **43** 801–26

Ambainis A, Bach E, Nayak A, Vishwanath A and Watrous J 2001 One-dimensional quantum walks *Proceedings of the Thirty-Third Annual ACM Symposium on Theory of Computing (New York)* (ACM) pp 37–49

Andréys S, Joye A and Raquépas R 2021 Fermionic walkers driven out of equilibrium *J. Statist. Phys.* **184** 33 Id/No 14

Annabestani M, Akhtarshenas S J and Abolhassani M R 2016 Incoherent tunneling effects in a one-dimensional quantum walk *J. Phys. A: Math. Theor.* **49** 19 Id/No 115 301

Apers S, Chakraborty S, Novo L and Roland J 2022 Quadratic speedup for spatial search by continuous-time quantum walk *Phys. Rev. Lett.* **129** Paper No. 160 502, 6

Arnault P and Cedzich C 2022 A single-particle framework for unitary lattice gauge theory in discrete time *New J. Phys.* **24** 123031

Arnault P and Debbasch F 2016 Landau levels for discrete-time quantum walks in artificial magnetic fields *Phys.* A **443** 179–91

Arnault P and Debbasch F 2017 Quantum walks and gravitational waves *Ann. Phys.* **383** 645–61

Arrighi P 2019 An overview of quantum cellular automata *Natural Computing* **18** 885–99

Arrighi P and Facchini S 2017 Quantum walking in curved spacetime: (3.1) dimensions and beyond *Quantum Inform. Compu.* **17** 810–24

Arrighi P, Facchini S and Forets M 2016 Quantum walking in curved spacetime *Quantum Inf. Process.* **15** 3467–86

Asahara K, Funakawa D, Seki M and Tanaka Y 2021 An index theorem for one-dimensional gapless non-unitary quantum walks *Quantum Inf. Process.* **20** 26 Id/No 287

Aso T and Maeda M 2021 A remark on uniform Strichartz estimate for quantum walks on 1D lattice *Yokohama Math.* J **67** 1–8

Attal S, Petruccione F and Sinayskiy I 2012 Open quantum walks on graphs *Phys. Lett.* A **376** 1545–8

Azbel' M Y 1973 Random two-component one-dimensional Ising model for heteropolymer melting *Phys. Rev. Lett.* **31** 589–92

Azbel' M Y 1995 Universality in a DNA statistical structure *Phys. Rev. Lett.* **75** 168–71

Bai F-l, Liu Y-z and Wang T-m 2007 A representation of DNA primary sequences by random walk *Math. Biosci.* **209** 282–91

Barkhofen S, Lorz L, Nitsche T, Silberhorn C and Schomerus H 2018 Supersymmetric polarization anomaly in photonic discrete-time quantum walks *Phys. Rev. Lett.* **121** 260501

Belton A C R and Wills S J 2015 An algebraic construction of quantum flows with unbounded generators *Annales de l'Institut Henri Poincaré. Probabilités et Statistiques* **51** 349–75

Bepari K, Malik S, Spannowsky M and Williams S 2022 Quantum walk approach to simulating parton showers *Phys. Rev.* D **106** 13 Paper No. 056 002, 13

Berger J A, Mitra S K, Carli M and Neri A 2004 Visualization and analysis of DNA sequences using DNA walks *J. Franklin Inst.* **341** 37–53

Bernaola-Galván P, Carpena P, Gómez-Martín C and Oliver J L 2023 Compositional structure of the genome: a review *Biology* **12** 849

Bezerra G A, Lugáo P H G and Portugal R 2021 *Phys. Rev.* A **103** 10 Paper No. 062 202, 10

Bhandari H and Durganandini P 2023 Long time dynamics of a single-particle extended quantum walk on a one-dimensional lattice with complex hoppings: a generalized hydrodynamic description *Quantum Inf. Process.* **22** 31 Paper No. 43

Bisio A, D'Ariano G M, Perinotti P and Tosini A 2015 Weyl, Dirac and Maxwell quantum cellular automata *Found. Phys.* **45** 1203–21

Boettcher S and Pughe-Sanford J L 2018 Renormalization of discrete-time quantum walks with a non-Grover coin *J. Stat. Mech: Theory Exp.* **2018** 19 Id/No 033 103

Bogoliubov N 2019 Continuous-time multidimensional walks as an integrable model *J. Math. Sci. (New York)* **238** 769–78

Bogoliubov N and Malyshev C 2017 Multi-dimensional random walks and integrable phase models *J. Math. Sci. (New York)* **224** 199–213

Böttcher L and Porter M A 2021 Classical and quantum random-walk centrality measures in multilayer networks *SIAM J. Appl. Math.* **81** 2704–24

Bourne C 2023 Index theory of chiral unitaries and split-step quantum walks *SIGMA Symmetry Integrability Geom. Methods Appl.* **19** Paper No. 053

Bressanini G, Benedetti C and Paris M G A 2022 Decoherence and classicalization of continuous-time quantum walks on graphs *Quantum Inf. Process.* **21** 317

Bringuier H 2017 Central limit theorem and large deviation principle for continuous time open quantum walks *Annales Henri Poincaré* **18** 3167–92

Brown L C, Martin W J and Wright D 2023 Continuous time quantum walks on graphs: group state transfer *Discrete Appl. Math.* **337** 68–80

Buarque A R C and Dias W S 2021 Probing coherence and noise tolerance in discrete-time quantum walks: unveiling self-focusing and breathing dynamics *Phys. Rev.* A **103** 7 Paper No. 042 213

Callison A, Chancellor N, Mintert F and Kendon V 2019 Finding spin glass ground states using quantum walks *New J. Phys.* **21** 123022

Candeloro A, Razzoli L, Cavazzoni S, Bordone P and Paris M G A 2020 Continuous-time quantum walks in the presence of a quadratic perturbation *Phys. Rev.* A **102** 19

Carbone R, Girotti F and Hernandez A M 2022 On a generalized central limit theorem and large deviations for homogeneous open quantum walks *J. Statist. Phys.* **188** 33 Id/No 8

Carbone R and Pautrat Y 2016 Open quantum random walks: reducibility, period, ergodic properties *Annales Henri Poincaré* **17** 99–135

Carteret H A, Richmond B and Temme N M 2005 Evanescence in coined quantum walks *J. Phys. A: Math. Gen.* **38** 8641–65

Carvalho S L, Guidi L F and Lardizabal C F 2017 Site recurrence of open and unitary quantum walks on the line *Quantum Inf. Process.* **16** 17

Cebrat S and Dudek M R 1998 The effect of DNA phase structure on DNA walks *European Physical Journal* B **3** 271–6

Cedzich C, Geib T, Stahl C, Velázquez L, Werner A H and Werner R F 2018 Complete homotopy invariants for translation invariant symmetric quantum walks on a chain *Quantum* **2** 95

Cedzich C, Geib T, Werner A H and Werner R F 2019 Quantum walks in external gauge fields *J. Math. Phys.* **60** 18

Cedzich C, Geib T, Werner A H and Werner R F 2021 Chiral Floquet systems and quantum walks at half-period *Annales Henri Poincaré* **22** 375–413

Cedzich C, Geib T and Werner R F 2022 An algorithm to factorize quantum walks into shift and coin operations *Lett. Math. Phys.* **112** 12 Id/No 85

Chakraborty S, Shaikh S H, Mandal S B, Ghosh R and Chakrabarti A 2019 A study and analysis of a discrete quantum walk-based hybrid clustering approach using d-regular bipartite graph and 1D lattice *Int. J. Quantum Inf.* **17** 27 Id/No 1 950 016

Chan A and Zhan H 2023 Pretty good state transfer in discrete-time quantum walks *J. Phys. A: Math. Theor.* **56** 25 Id/No 165 305

Chapon F and Defosseux M 2012 Quantum random walks and minors of Hermitian Brownian motion *Can. J. Math.* **64** 805–21

Chaudhury S S and Dutta S 2022 Quantum multi-secret sharing via trap codes and discrete quantum walks *Quantum Inf. Process.* **21** 380

Chaves R, Chagas B and Coutinho G 2023 Why and how to add direction to a quantum walk *Quantum Inf. Process.* **22** 15 Paper No. 41

Chaves R, Santos J and Chagas B 2023 Transport properties in directed quantum walks on the line *Quantum Inf. Process.* **22** 14 Id/No 144

Chen F, Zhang H, Chen S and Cheng W 2021 Novel two-party quantum private comparison via quantum walks on circle *Quantum Inf. Process.* **20** 178

Chen M, Ferro G M and Sornette D 2022 On the use of discrete-time quantum walks in decision theory *PLOS ONE* **17** e0273551

Chen T, Wang B and Zhang X 2017 Controlling probability transfer in the discrete-time quantum walk by modulating the symmetries *New J. Phys.* **19** 113049

Chen X and Lou X 2022 An efficient verifiable quantum secret sharing scheme via quantum walk teleportation *Int. J. Theor. Phys.* **61** 10 Paper No. 99

Chen Y, Meng X, Lu H and Dong H 2022 Engineering DNA walkers for bioanalysis: a review *Analytica chimica acta* **1209** 339339

Chen Y, Xiang Y, Yuan R and Chai Y 2014 A restriction enzyme-powered autonomous DNA walking machine: its application for a highly sensitive electrochemiluminescence assay of DNA *Nanoscale* **7** 981–6

Chiang C-F, Nagaj D and Wocjan P 2010 Efficient circuits for quantum walks Quantum *Inf. Comput.* **10** 420–34

Childs A M 2009 universal computation by quantum walk *Phys. Rev. Lett.* **102** 180501

Childs A M, Farhi E and Gutmann S 2002 An example of the difference between quantum and classical random walks *Quantum Information Processing* **1** 35–43

Childs A M and Goldstone J 2004 Spatial search by quantum walk *Phys. Rev.* A **70** 022314

Chisaki K, Konno N, Segawa E and Shikano Y 2011 Crossovers induced by discrete-time quantum walks *Quantum Inf. Comput.* **11** 741–60

Condon E A and Rozenberg G 2001 *DNA computing vol 2054 of Lecture Notes in Computer Science* Berlin Springer-Verlag Papers from the 6th International Workshop on DNA-based Computers (DNA 2000) held at the University of Leiden, Leiden, June 13–17, 2000

Costa P C S, Portugal R and de Melo F 2018 Quantum walks via quantum cellular automata *Quantum Inf. Process.* **17** 226

Coutinho G, Godsil C, Juliano E and van Bommel C M 2022 Quantum walks do not like bridges *Linear Algebra Appl* **652** 155–72

Crespi A, Osellame R, Ramponi R, Giovannetti V, Fazio R, Sansoni L, de Nicola F, Sciarrino F and Mataloni P 2013 Anderson localization of entangled photons in an integrated quantum walk *Nat. Photonics* **7** 322–8

Curado M, Escolano F, Hancock E R, Nourbakhsh F and Pelillo M 2015 Similarity analysis from limiting quantum walks *Similarity-Based Pattern Recognition. Third International Workshop, SIMBAD 2015, Copenhagen, Denmark, October 12–14, 2015. Proceedings (Cham)* (Springer) pp 38–53

Dai W, Yuan J and Li D 2018 Discrete-time quantum walk on the Cayley graph of the dihedral group *Quantum Inf. Process.* **17** 21 Id/No 330

Darázs Z and Kiss T 2013 Time evolution of continuous-time quantum walks on dynamical percolation graphs *J. Phys. A: Math. Theor.* **46** 14 Id/No 375 305

D'Ariano G M, Erba M, Perinotti P and Tosini A 2017 Virtually abelian quantum walks *J. Phys. A: Math. Theor.* **50** 18 Id/No 035 301

de J, León-Montiel R, Méndez V, Quiroz-Juárez M A, Ortega A, Benet L, Perez-Leija A and Busch K 2019 Two-particle quantum correlations in stochastically-coupled networks *New J. Phys.* **21** 053041

Debbasch F 2023 Minimal quantum walk simulation of Dirac fermions in curved space-times *Quantum Stud. Math. Found.* **10** 317–27

Dhahri A, Ko C K and Yoo H J 2019 Quantum markov chains associated with open quantum random walks *J. Statist. Phys.* **176** 1272–95

Di Molfetta G, Honter L, Luo B B, Wada T and Shikano Y 2015 Massless Dirac equation from Fibonacci discrete-time quantum walk *Quantum Studies: Mathematics and Foundations* **2** 243–52

Didi A and Barkai E 2022 Measurement-induced quantum walks *Phys. Rev.* E **105** 054108

Doliskani J 2023 How to sample from the limiting distribution of a continuous-time quantum walk *IEEE Trans. Inf. Theory* 1

Duan L 2022 Quantum walk on the Bloch sphere *Phys. Rev.* A **105** 6 Paper No. 042 215

Duplij D and Duplij S 2000 Symmetry analysis of genetic code and determinative degree *Biophysical Bull. Kharkov Univ.* **488** 60–70

Duplij D and Duplij S 2001 Determinative degree and nucleotide content of DNA strands *Biophysical Bull. Kharkov Univ.* **525** 86–92

Duplij D and Duplij S 2005 DNA sequence representation by trianders and determinative degree of nucleotides *J. Zhejiang Univ. Sci.* B **6** 743–55

Duplij D, Duplij S and Chashchin N 2000 Symmetric properties of genetic code *Biopolymers and Cell* **16** 449–54

D'Ariano G M, Mosco N, Perinotti P and Tosini A 2016 Discrete time dirac quantum walk in 3 +1 dimensions *Entropy-switz.* **18** 228

Li E W and Vanden-Eijnden E E 2019 *Applied stochastic analysis vol 199 of Graduate Studies in Mathematics* (Providence, RI: American Mathematical Society)

El-Latif A A A, Iliyasu A M and Abd-El-Atty B 2021 An efficient visually meaningful quantum walks-based encryption scheme for secure data transmission on iot and smart applications *Mathematics* **9** 3131

Emms D, Severini S, Wilson R C and Hancock E R 2009 Coined quantum walks lift the cospectrality of graphs and trees *Pattern Recogn* **42** 1988–2002

Emms D, Wilson R C and Hancock E R 2009 Graph matching using the interference of discrete-time quantum walks *Image Vis. Comput.* **27** 934–49

Endo T, Konno N and Obuse H 2020 Relation between two-phase quantum walks and the topological invariant *Yokohama Math.* J **66** 1–59

Endo T, Konno N, Segawa E and Takei M 2014 A one-dimensional Hadamard walk with one defect Yokohama *Math.* J **60** 49–90

Ewens W J 2001 Mathematics and the human genome project *Math. Sci.* **26** 1–12

Faccin M, Johnson T, Biamonte J, Kais S and Migdał P 2013 Degree distribution in quantum walks on complex networks *Phys. Rev.* X **3** 041007

Fan H, Wu Y, Huang T, Hong N, Cui H, Wei G, Liao F and Zhang J 2022 An electrochemical DNA sensor based on an integrated and automated DNA walker *Bioelectrochemistry (Amsterdam, Netherlands)* **147** 108198

Fedichkin L E and Meshchaninov F P 2021 Analysis and applications of quantum walks *J. Math. Sci. (New York)* **252** 104–15

Feldman E and Hillery M 2007 Modifying quantum walks: a scattering theory approach *J. Phys. A: Math. Theor.* **40** 11343–59

Feng Y, Shi R, Shi J, Zhou J and Guo Y 2019 Arbitrated quantum signature scheme with quantum walk-based teleportation *Quantum Inf. Process.* **18** 21 Id/No 154

Freslon A 2018 Quantum reflections, random walks and cut-off *Int. J. Math.* **29** 23 Id/No 1 850 101

Frigerio M and Paris M G A 2023 Swift chiral quantum walks *Linear Algebra Appl* **673** 28–45

Fuda T, Funakawa D and Suzuki A 2017 Localization of a multi-dimensional quantum walk with one defect *Quantum Inf. Process.* **16** 24 Id/No 203

Fuss I, White L, Sherman P and Naguleswaran S 2007 *An Analytic Solution for One-Dimensional Quantum Walks* preprint University of Adelaide Adelaide 23 p. (arxiv:0705.0077)

Gawron P, Kurzyk D and Puchała Z 2013 A model for quantum queue *Int. J. Quantum Inf.* **11** 15 Id/No 1 350 023

Georgopoulos K, Emary C and Zuliani P 2021 Comparison of quantum-walk implementations on noisy intermediate-scale quantum computers *Phys. Rev.* A **103** 10 Paper No. 022 408

Gettrick M M 2010 One dimensional quantum walks with memory *Quantum Inf. Comput.* **10** 509–24

Gilbert D, Heiner M and Rohr C 2018 Petri-net-based 2D design of DNA walker circuits *Nat. Comput.* **17** 161–82

Gillespie D T 1977 Exact stochastic simulation of coupled chemical reactions *J. Phys. Chem.* **81** 2340–61

Giordani T *et al* 2019 Experimental engineering of arbitrary qudit states with discrete-time quantum walks *Phys. Rev. Lett.* **122** 020503

Giri P R 2023 Quantum walk search on a two-dimensional grid with extra edges *Int. J. Theor. Phys.* **62** 12 Paper No. 121

Glos A, Krawiec A, Kukulski R and Puchała Z 2018 Vertices cannot be hidden from quantum spatial search for almost all random graphs *Quantum Inf. Process.* **17** 15 Id/No 81

Godoy S and Fujita S 1992 A quantum random-walk model for tunneling diffusion in a 1D lattice. a quantum correction to Fick's law *J. Chem. Phys.* **97** 5148–54

Godsil C 2013 Average mixing of continuous quantum walks *J. Combin. Theory Ser.* A **120** 1649–62

Godsil C, Guo K and Myklebust T G J 2017 *J. Combin.* **24** 6 Paper No. 4.16

Gottlieb A D 2005 Convergence of continuous time quantum walks on the line *Phys. Rev. E Stat. Nonlin. Soft Matter Phys.* **72** 047102

Groenland K and Bannink T 2017 Searching faster with quantum random walks *Nieuw Arch. Wiskd.* **18** 249–54

Grössing G and Zeilinger A 1988 Quantum cellular automata *Complex Syst.* **2** 197–208

Grudka A, Kurzyński P, Polak T P, Sajna A S, Wójcik J and Wójcik A 2023 Complementarity in quantum walks *J. Phys.* A **56** 22 Paper No. 275 303

Grünbaum F A, Vinet L and Zhedanov A 2013 Birth and death processes and quantum spin chains *J. Math. Phys.* **54** 12

Guan J, Wang Q and Ying M 2021a An HHL-based algorithm for computing hitting probabilities of quantum walks *Quantum Inform. Compu.* **21** 395–408

Guan J, Wang Q and Ying M 2021b quantum walks *Quantum Inf. Comput.* **21** 395–408

Haimovich A D, Byrne B, Ramaswamy R and Welsh W J 2006 Wavelet analysis of DNA walks *J. Comput. Biol.* **13** 1289–98

Hamilton C S, Barkhofen S, Sansoni L, Jex I and Silberhorn C 2016 Driven discrete time quantum walks *New J. Phys.* **18** 073008

Hamilton C S, Kruse R, Sansoni L, Silberhorn C and Jex I 2014 Driven quantum walks *Phys. Rev. Lett.* **113** 083602

Hamza E and Joye A 2014 Spectral transition for random quantum walks on trees *Comm. Math. Phys.* **326** 415–39

Han Q, Kou Y, Bai N and Wang H 2022 The continuous-time quantum walk on some graphs based on the view of quantum probability *Int. J. Quantum Inf.* **20** 13 Id/No 2 250 015

Han Q, Kou Y, Wang H and Bai N 2023 An investigation of continuous-time quantum walk on hypercube in view of Cartesian product structure *Int. J. Quantum Inf.* **21** 12 Paper No. 2 350 012

Han Q, Lu Z, Han Y and Chen Z 2022 Discrete-time quantum random walks on the N-ary tree *Chinese J. Appl. Probab. Statist* **38** 86–98

Han X, Feng L, Li Y, Zhang L, Song J and Zhang Y 2019 Experimental observations of 1D quantum walks in a limited region *Quantum Inf. Process.* **18** 85

Hanaoka R and Konno N 2021 Return probability and self-similarity of the Riesz walk *Quantum Inform. Compu.* **21** 409–22

He Z, Huang Z, Li L and Situ H 2017 Coherence of one-dimensional quantum walk on cycles *Quantum Inf. Process.* **16** 14 Id/No 271

Hegde A S and Chandrashekar C M 2022 Characterization of anomalous diffusion in one-dimensional quantum walks *J. Phys. A: Math. Theor.* **55** 14 Id/No 234 006

Herrman R and Humble T S 2019 Continuous-time quantum walks on dynamic graphs *Phys. Rev.* A **100** 13

Herrman R and Wong T G 2022 Simplifying continuous-time quantum walks on dynamic graphs *Quantum Inf. Process.* **21** 29 Id/No 54

Higuchi Y, Konno N, Sato I and Segawa E 2012 Quantum graph walks I: mapping to quantum walks *Yokohama Math.* J **59** 33–56

Higuchi Y, Konno N, Sato I and Segawa E 2013 A note on the discrete-time evolutions of quantum walk on a graph *J. Ind. Math.* **5** 103–9

Higuchi Y, Portugal R, Sato I and Segawa E 2019 Eigenbasis of the evolution operator of 2-tessellable quantum walks *Linear Algebra Appl.* **583** 257–81

Hillery M, Bergou J and Feldman E 2003 Quantum walks based on an interferometric analogy *Phys. Rev.* A **68** 032314

Hines A P and Stamp P C E 2007 Quantum walks, quantum gates and quantum computers *Phys. Rev.* A **75** 062321

Ho C-L, Ide Y, Konno N, Segawa E and Takumi K 2018 A spectral analysis of discrete-time quantum walks related to the birth and death chains *J. Statist. Phys.* **171** 207–19

Hou P, Shang T, Zhang Y, Tang Y and Liu J 2023 Quantum hash function based on controlled alternate lively quantum walks *Sci. Rep.* **13** 5887

Hou Z, Tang J-F, Shang J, Zhu H, Li J, Yuan Y, Wu K-D, Xiang G-Y, Li C-F and Guo G-C 2018 Deterministic realization of collective measurements via photonic quantum walks *Nat. Commun.* **9** 1414

Høyer P and Yu Z 2020 Analysis of lackadaisical quantum walks *Quantum Inf. Comput.* **20** 1137–52

Huerta Alderete C, Singh S, Nguyen N H, Zhu D, Balu R, Monroe C, Chandrashekar C M and Linke N M 2020 Quantum walks and dirac cellular automata on a programmable trapped-ion quantum computer *Nat. Commun.* **11** 3720

Iche G and Nozieres P 1978 Quantum Brownian motion of a heavy particle: an adiabatic expansion *Physica A Stat. Mech. Appl.* **91** 485–506

Ide Y, Konno N and Machida T 2011 Entanglement for discrete-time quantum walks on the line *Quant. Inf. Comput.* **11** 855–66

Jadczyk A 2022 Random walk on quantum blobs *Open Syst. Inf. Dyn.* **29** 25 Paper No. 2 250 013

Jafarizadeh M A and Sufiani R 2007 Investigation of continuous-time quantum walks via spectral analysis and Laplace transform *Int. J. Quantum Inf.* **5** 575–96

Jay G, Debbasch F and Wang J 2020 A systematic method to building dirac quantum walks coupled to electromagnetic fields *Quantum Inf. Process.* **19** 422

Jayakody M N, Nanayakkara A and Cohen E 2021 Analysis of decoherence in linear and cyclic quantum walks *Optics* **2** 236–50

Ju L, Yang M, Paunković N, Chu W-J and Cao Z-L 2019 Creating photonic GHZ and W states via quantum walk *Quantum Inf. Process.* **18** 12 Id/No 176

Jung C, Allen P B and Ellington A D 2016 A stochastic DNA walker that traverses a microparticle surface *Nat. Nanotechnol.* **11** 157–63

Kadian K, Garhwal S and Kumar A 2021 Quantum walk and its application domains: a systematic review *Comput. Sci. Rev.* **41** 52 Id/No 100 419

Kang Y and Wang C 2014 Itô formula for one-dimensional continuous-time quantum random walk *Phys.* A **414** 154–62

Kang Y B 2023 Markov properties of partially open quantum random walks *J. Math. Phys.* **64** 12 Id/No 033 301

Kargin V 2010 Continuous-time quantum walk on integer lattices and homogeneous trees *J. Statist. Phys.* **140** 393–408

Kaszlikowski D and Kurzyński P 2021 A little bit of classical magic to achieve (super-)quantum speedup *Found. Phys.* **51** 14 Id/No 55

Katayama H, Hatakenaka N and Fujii T 2020a Floquet-engineered quantum walks *Sci. Rep.* **10** 17544

Katayama H, Hatakenaka N and Fujii T 2020b Theoretical studies on quantum walks with a time-varying coin *Proc. 9th Int. Conf. Quantum Simulation and Quantum Walks, QSQW 2020, Marseille, France, January 20–24, 2020 (Waterloo)* (Open Publishing Association (OPA))

Kendon V 2006a Quantum walks on general graphs *Int. J. Quantum Inf.* **4** 791–805

Kendon V 2006b A random walk approach to quantum algorithms *Phil. Trans. R. Soc. A (2006)* **364** 3407–22

Kendon V 2020 How to compute using quantum walks *EPTCS* **315** 1–17

Kendon V and Maloyer O 2008 Optimal computation with non-unitary quantum walks *Theor. Comput. Sci.* **394** 187–96

Kendon V and Tregenna B 2003 Decoherence can be useful in quantum walks *Phys. Rev.* A **67** 042315

Khadiev K, Nahimovs N and Santos R A M 2018 On the probability of finding marked connected components using quantum walks *Lobachevskii J. Math.* **39** 1016–23

Kieferová M and Nagaj D 2012 Quantum walks on necklaces and mixing *Int. J. Quantum Inf.* **10** 16

Kitagawa T, Rudner M S, Berg E and Demler E 2010 Exploring topological phases with quantum walks *Phys. Rev.* A **82** 033429

Kiumi C 2021 A new type of quantum walks based on decomposing quantum states *Quantum Inf. Comput.* **21** 541–56

Kiumi C 2022a Localization in quantum walks with periodically arranged coin matrices *Int. J. Quantum Inf.* **20** 14

Kiumi C 2022b Localization of space-inhomogeneous three-state quantum walks *J. Phys. A: Math. Theor.* **55** 14

Kiumi C, Konno N and Tamura S 2022 Return probability of quantum and correlated random walks *Entropy (Basel, Switzerland)* **24** 584

Komatsu T, Konno N, Morioka H and Segawa E 2022 Asymptotic properties of generalized eigenfunctions for multi-dimensional quantum walks *Annales Henri Poincaré 23 1693-1724*

Konno N 2006 Continuous-time quantum walks on trees in quantum probability theory *Infin. Dimens. Anal. Quantum Probab. Relat. Top* **9** 287–97

Konno N 2009 Limit theorems and absorption problems for one-dimensional correlated random walks *Stoch. Models* **25** 28–49

Konno N 2015 Quaternionic quantum walks *Quantum Stud.: Math. Found.* **2** 63–76

Konno N 2020 Quantum walks *Sugaku Expositions* **33** 135–58

Konno N 2022 An analogue of the Riemann Hypothesis via quantum walks Quantum *Stud. Math. Found* **9** 367–79

Konno N and Machida T 2010 Limit theorems for quantum walks with memory *Quant. Inf. Comp* **10** 1004–17

Konno N, Matsue K, Mitsuhashi H and Sato I 2017 Quaternionic quantum walks of szegedy type and zeta functions of graphs *Quantum Inf. Comput.* **17** 1349–71

Konno N, Mitsuhashi H and Sato I 2016 The discrete-time quaternionic quantum walk on a graph *Quantum Inf. Process.* **15** 651–73

Konno N, Mitsuhashi H and Sato I 2017 Quaternionic Grover walks and zeta functions of graphs with loops *Graphs Combin* **33** 1419–32

Konno N, Portugal R, Sato I and Segawa E 2018 Partition-based discrete-time quantum walks *Quantum Inf. Process.* **17** 100

Kubota S 2023 Combinatorial necessary conditions for regular graphs to induce periodic quantum walks *Linear Algebra Appl.* **673** 259–79

Kubota S, Saito K and Yoshie Y 2022 A new type of spectral mapping theorem for quantum walks with a moving shift on graphs *Quantum Inf. Process.* **21** 30 Id/No 159

Kubota S, Segawa E, Taniguchi T and Yoshie Y 2019 A quantum walk induced by Hoffman graphs and its periodicity *Linear Algebra Appl.* **579** 217–36

Kuklinski P and Kon M A 2018 Absorption probabilities of quantum walks *Quantum Inf. Process.* **17** 263

Kuriki S, Nirjhor M S A and Ohno H 2021 Parameterization of quantum walks on cycles *Quantum Inf. Process.* **20** 28

Kurzyński P and Wójcik A 2011 Discrete-time quantum walk approach to state transfer *Phys. Rev.* A **83** 062315

Kurzyński P and Wójcik A 2013 Quantum walk as a generalized measuring device *Phys. Rev. Lett.* **110** 200404

Lagro M, Yang W-S and Xiong S 2017 A Perron-Frobenius type of theorem for quantum operations *J. Statist. Phys.* **169** 38–62

Lehman L, Zatloukal V, Brennen G K, Pachos J K and Wang Z 2011 Quantum walks with non-abelian anyons *Phys. Rev. Lett.* **106** 230404

Li B-h 2004 Describing DNA sequences by 2-dimensional walks *Acta Anal. Funct. Appl.* **6** 1–4

Li D, Gettrick M M, Zhang W-W and Zhang K-J 2014 One dimensional lazy quantum walks and occupancy rate *Chinese Physics B, March 2015, vol 24 p 050 305-1-050305-8* **24**

Li D, Mc Gettrick M, Yang Y-G, Xu J and Wang Y 2020 2020 Quantum walks with memory provided by parity of memory *Int. J. Theor. Phys.* **59** 1934–43

Li D, Yang Y-G, Bi J-L, Yuan J-B and Xu J 2018 Controlled alternate quantum walks based quantum hash function *Sci. Rep.* **8** 225

Li D, Zhang J, Guo F, Huang W, Wen Q and Chen H 2013 Discrete-time interacting quantum walks and quantum hash schemes *Quantum Inf. Process.* **12** 1501–13

Li H, Chen X, Wang Y, Hou Y and Li J 2019 A new kind of flexible quantum teleportation of an arbitrary multi-qubit state by multi-walker quantum walks *Quantum Inf. Process.* **18** 266

Li H, Li J, Xiang N, Zheng Y, Yang Y and Naseri M 2019 A new kind of universal and flexible quantum information splitting scheme with multi-coin quantum walks *Quantum Inf. Process.* **18** 316

Li H-J, Li J and Chen X 2022 Generalized quantum teleportation of shared quantum secret: a coined quantum-walk approach *Quantum Inf. Process.* **387** 21 Paper No. 387

Li Q, He Y and Jiang J 2011 A hybrid classical-quantum clustering algorithm based on quantum walks *Quantum Inf. Process.* **10** 13–26

Li X, Wu M, Chen H and Liu Z 2021 Algorithms for finding the maximum clique based on continuous time quantum walks *Quantum Inf. Comput.* **21** 59–79

Li X-Y, Chang Y, Zhang S-B, Dai J-Q and Zheng T 2020 2020 Quantum blind signature scheme based on quantum walk *Int. J. Theor. Phys.* **59** 2059–73

Li Z-J and Wang J B 2015 An analytical study of quantum walk through glued-tree graphs *J. Phys. A: Math. Theor.* **58** 15 Id/No 355 301

Liang W, Yan F, Iliyasu A M, Salama A S and Hirota K 2022 A simplified quantum walk model for predicting missing links of complex networks *Entropy* **24** 13 Paper No. 1547

Liu G, Li W, Fan X, Li Z, Wang Y and Ma H 2022 An image encryption algorithm based on discrete-time alternating quantum walk and advanced encryption standard *Entropy* **24** 16 Paper No. 608

Liu Y, Su W J and Li T 2023 On quantum speedups for nonconvex optimization via quantum tunneling walks *Quantum* **7** 1030

Liu Y, Yuan J, Duan B and Li D 2018 Quantum walks on regular uniform hypergraphs *Sci. Rep.* **8** 9548

Lo P, Rajaram S, Schepens D, Sullivan D, Tamon C and Ward J 2006 Mixing of quantum walk on circulant bunkbeds *Quantum Inform. Compu.* **6** 370–81

Lobry J R 1996 A simple vectorial representation of DNA sequences for the detection of replication origins in bacteria *Biochimie* **78** 323–6

Loebens N 2023 Site recurrence for continuous-time open quantum walks on the line *Quantum Inf. Comput.* **23** 577–602

Lomoc F, Boette A P, Canosa N and Rossignoli R 2022 History states of one-dimensional quantum walks *Phys. Rev.* A **106** 12 Paper No. 062 215

Lv Q, Cao P, Cohen E, Li K and Shenker S 2002 Search and replication in unstructured peer-to-peer networks *Proc. 16th Int. Conf. Supercomputing: ACM* (New York: ACM) pp 84–95

Machida T 2013a Limit theorems for the interference terms of discrete-time quantum walks on the line *Quantum Inf. Comput.* **13** 661–71

Machida T 2013b A quantum walk with a delocalized initial state: contribution from a coin-flip operator *Int. J. Quantum Inf.* **11** 13 Id/No 1 350 053

Machida T 2018 A limit theorem for a splitting distribution of a quantum walk *Int. J. Quantum Inf.* **16** 11 Id/No 1 850 023

Machida T 2021 A limit distribution for a quantum walk driven by a five-diagonal unitary matrix *Quantum Inform. Compu.* **21** 19–36

Machida T and Grünbaum F A 2018 Some limit laws for quantum walks with applications to a version of the parrondo paradox *Quantum Inf. Process.* **17** 241

Maeda M 2022 Asymptotic stability of small bound state of nonlinear quantum walks *Phys.* D **439** 14

Maeda M, Sasaki H, Segawa E, Suzuki A and Suzuki K 2018 Scattering and inverse scattering for nonlinear quantum walks *Discrete Contin. Dyn. Syst.* **38** 3687–703

Maeda M and Suzuki A 2020 Continuous limits of linear and nonlinear quantum walks *Rev. Math. Phys.* **32** 20

Magniez F, Nayak A, Richter P C and Santha M 2012 On the hitting times of quantum versus random walks *Algorithmica* **63** 91–116

Magniez F, Nayak A, Roland J and Santha M 2011 Search via quantum walk *SIAM J. Comput.* **40** 142–64

Manouchehri K and Wang J B 2007 Continuous-time quantum random walks require discrete space *J. Phys. A: Math. Theor.* **40** 13773–85

Manouchehri K and Wang J B 2008 Quantum walks in an array of quantum dots *J. Phys. A: Math. Theor.* **41** 12

Markiewicz M, Karczewski M and Kurzynski P 2021 Borromean states in discrete-time quantum walks *Quantum* **5** 523

Marsh S and Wang J B 2019 A quantum walk-assisted approximate algorithm for bounded NP optimisation problems *Quantum Inf. Process.* **18** 18

Martínez-Martínez I and Sánchez-Burillo E 2016 Quantum stochastic walks on networks for decision-making *Sci. Rep.* **6** 23812

Matsue K, Matsuoka L, Ogurisu O and Segawa E 2019 Resonant-tunneling in discrete-time quantum walk *Quantum Studies: Mathematics and Foundations* **6** 35–44

Matsue K, Ogurisu O and Segawa E 2017 A note on the spectral mapping theorem of quantum walk models *Interdiscip. Inf. Sci. (IIS)* **23** 105–14

Matsuzawa Y 2020 An index theorem for split-step quantum walks *Quantum Inf. Process.* **19** 227

Matwiejew E and Wang J 2021 QSW_MPI: a framework for parallel simulation of quantum stochastic walks *Comput. Phys. Commun.* **260** 12

Mendes C V C, Almeida G M A, Lyra M L and de Moura F A B F 2019 Localization-delocalization transition in discrete-time quantum walks with long-range correlated disorder *Phys. Rev.* E **99** 022117

Mendes C V C, Almeida G M A, Lyra M L and de Moura F A B F 2021 Localization properties of a discrete-time 1D quantum walk with generalized exponential correlated disorder *Phys. Lett.* A **394** 6

Mi D, Song H S and Hou B Y 1999 General random walk in one dimension and its application to DNA transcription process *Commun. Theor. Phys.* **31** 637–40

Miki H, Tsujimoto S and Vinet L 2022 Classical and quantum walks on paths associated with exceptional Krawtchouk polynomials *J. Math. Phys.* **63** 16

Mochizuki K, Bessho T, Sato M and Obuse H 2020 Topological quantum walk with discrete time-glide symmetry *Phys. Rev.* B **102** 035418

Monterde H 2023 Sedentariness in quantum walks *Quantum Inf. Process.* **22** Paper No. 273

Moore C and Russell A 2002 Quantum walks on the hypercube *Randomization and Approximation Techniques in Computer Science* ed J D P Rolim and S Vadhan (Berlin-Heidelberg: Springer) pp 164–178

Muraleedharan G, Miyake A and Deutsch I H 2019 Quantum computational supremacy in the sampling of bosonic random walkers on a one-dimensional lattice *New J. Phys.* **21** 055003

Mylläri A, Salakoski T and Pasechnik A 2005 On the visualization of the DNA sequence and its nucleotide content *ACM SIGSAM Bulletin* **39** 131–5

Mülken O and Blumen A 2006 Continuous-time quantum walks in phase space *Phys. Rev.* A **73** 012105

Mülken O, Pernice V and Blumen A 2008 universal behavior of quantum walks with long-range steps *Phys. Rev. E Stat. Nonlin. Soft Matter Phys.* **77** 021117

Nahimovs N, Santos R A M and Khadiev K R 2019 Adjacent vertices can be hard to find by quantum walks *Moscow Univ. Comput. Math. Cybernet.* **43** 32–9

Narimatsu A 2021 Localization does not occur for the Fourier walk on the multi-dimensional lattice *Quantum Inform. Compu.* **21** 387–94

Narimatsu A, Ohno H and Wada K 2021 Unitary equivalence classes of split-step quantum walks *Quantum Inf. Process.* **20** 368
Nayak A and Vishwanath A 2000 *Quantum Walk on the Line* (arXiv e-prints quant–ph/0010117)
Nee S 1992 Uncorrelated DNA walks *Nature* **357** 450
Nguyen D T, Nguyen T A, Khrapko R, Nolan D A and Borrelli N F 2020 Quantum walks in periodic and quasiperiodic fibonacci fibers *Sci. Rep.* **10** 7156
Nitsche T, Barkhofen S, Kruse R, Sansoni L, Štefaňák M, Gábris A, Potoček V, Kiss T, Jex I and Silberhorn C 2018 Probing measurement-induced effects in quantum walks via recurrence *Sci. Adv.* **4** eaar6444
Noh J D and Rieger H 2004 Random walks on complex networks *Phys. Rev. Lett.* **92** 118701
Ohya S 2023 Quantum walk on orbit spaces *Phys. Rev.* A **107** Paper No. 062 202
Omanakuttan S and Lakshminarayan A 2018 Quantum walk on a toral phase space *J. Phys. A: Math. Theor.* **51** 21 Id/No 385 306
Omanakuttan S and Lakshminarayan A 2019 Out-of-time-ordered correlators and quantum walks *Phys. Rev.* E **99** 062128
Ortega S A and Martin-Delgado M A 2023 Discrete-time semiclassical Szegedy quantum walks *Phys.* A **625** 18 Id/No 129 021
Orthey A C 2017 Asymptotic entanglement in quantum walks from delocalized initial states *Quantum Inf. Process.* **16** 16 Id/No 224
Ouvry S and Polychronakos A P 2022 Algebraic area enumeration for open lattice walks *J. Phys. A: Math. Theor.* **55** 15 Id/No 485 005
Paxia S, Rudra A, Zhou Y and Mishra B 2002 A random walk down the genomes: DNA evolution in valis *Computer* **35** 73–9
Peng C K, Buldyrev S V, Goldberger A L, Havlin S, Sciortino F, Simons M and Stanley H E 1992 Fractal landscape analysis of DNA walks *Physica A Stat. Mech. Appl.* **191** 25–9
Philipp P and Portugal R 2017 *Quantum Inf. Process.* **16** 15 Id/No 14
Ponomarev A L and Sachs R K 2001 Radiation breakage of DNA: a model based on random-walk chromatin structure *J. Math. Biol.* **43** 356–76
Portugal R 2013 *Quantum walks and search algorithms* 2nd edn (New York: Springer)
Portugal R 2016 Staggered quantum walks on graphs *Phys. Rev.* A **93** 13
Portugal R, de Oliveira M C and Moqadam J K 2017 Staggered quantum walks with Hamiltonians *Phys. Rev.* A **95** 6
Portugal R, Santos R A M, Fernandes T D and Gonçalves D N 2016 The staggered quantum walk model *Quantum Inf. Process.* **15** 85–101
Portugal R and Segawa E 2017 Connecting coined quantum walks with Szegedy's model *Interdiscip. Inf. Sci. (IIS)* **23** 119–25
Prabhu P and Brun T A 2022 Influence of coin symmetry on infinite hitting times in quantum walks *Phys. Rev.* A **105** 8 Paper No. 032 206
Qi F, Wang Y F, Ma Q Y and Zheng W H 2016 Experimentally simulating quantum walks with self-collimated light *Sci. Rep.* **6** 28610
Qiang X, Loke T, Montanaro A, Aungskunsiri K, Zhou X, O'Brien J L, Wang J B and Matthews J C F 2016 Efficient quantum walk on a quantum processor *Nat. Commun.* **7** 1–6
Qu D, Marsh S, Wang K, Xiao L, Wang J and Xue P 2022 Deterministic search on star graphs via quantum walks *Phys. Rev. Lett.* **128** 050501
Reitzner D, Nagaj D and Bužek V 2011 Quantum walks *Acta Phys. Slovaca* **61** 603–725

Rigovacca L and Di Franco C 2016 Two-walker discrete-time quantum walks on the line with percolation *Sci. Rep.* **6** 22052

Rodrigues J, Paunković N and Mateus P 2017 A simulator for discrete quantum walks on lattices *Internat. J. Modern Phys.* C **28** 27

Rosas A, Nogueira E and Fontanari J F 2002 Multifractal analysis of DNA walks and trails *Phys. Rev.* E **66** 061906

Rousseva J and Kovchegov Y 2017 On alternating quantum walks *Phys.* A **470** 309–20

Rudner M S and Levitov L S 2009 Topological transition in a non-hermitian quantum walk *Phys. Rev. Lett.* **102** 065703

Ruiz-Ortiz M A, Martín-González E M, Santiago-Alarcon D and Venegas-Andraca S E 2023 A new definition of hitting time and an embedded Markov chain in continuous-time quantum walks *Quantum Inf. Process.* **22** 224

Sadowski P and Pawela L U 2016 Central limit theorem for reducible and irreducible open quantum walks *Quantum Inf. Process.* **15** 2725–43

Sajna A S, Polak T P, Wójcik A and Kurzyński P 2019 Attractors and asymptotic dynamics of open discrete-time quantum walks on cycles *Phys. Rev.* A **100** 9

Sako H 2021 Convergence theorems on multi-dimensional homogeneous quantum walks *Quantum Inf. Process.* **20** 1–24

Salimi S 2008 Study of continuous-time quantum walks on quotient graphs via quantum probability theory *Int. J. Quantum Inf.* **6** 945–57

Salimi S and Sorouri A 2010 Pseudo-Hermitian continuous-time quantum walks *J. Phys. A: Math. Theor.* **43** Id/No 275 304

Sansoni L, Sciarrino F, Vallone G, Mataloni P, Crespi A, Ramponi R and Osellame R 2012 Two-particle bosonic-fermionic quantum walk via integrated photonics *Phys. Rev. Lett.* **108** 010502

Santos R A M 2016 Szegedy's quantum walk with queries *Quantum Inf. Process.* **15** 4461–75

Schmitz H, Matjeschk R, Schneider C, Glueckert J, Enderlein M, Huber T and Schaetz T 2009 Quantum walk of a trapped ion in phase space *Phys. Rev. Lett.* **103** 090504

Schreiber A, Cassemiro K N, Potoček V, Gábris A, Jex I and Silberhorn C 2011 Decoherence and disorder in quantum walks: from ballistic spread to localization *Phys. Rev. Lett.* **106** 180403

Schwinger J 1961 Brownian motion of a quantum oscillator *J. Math. Phys.* **2** 407–32

Sett A, Pan H, Falloon P E and Wang J 2019 Zero transfer in continuous-time quantum walks *Quantum Inf. Process.* **18** 159

Sharma R and Boettcher S 2022 Transport and localization in quantum walks on a random hierarchy of barriers *J. Phys. A: Math. Theor.* **55** 15

Shenvi N, Kempe J and Whaley K B 2003 Quantum random-walk search algorithm *Phys. Rev.* A **67** 052307

Shi W-M, Bai M-X, Zhou Y-H and Yang Y-G 2023 Controlled quantum teleportation based on quantum walks *Quantum Inf. Process.* **22** 36

Shikano Y and Katsura H 2010 Localization and fractality in inhomogeneous quantum walks with self-duality *Phys. Rev. E Stat. Nonlin. Soft Matter Phys.* **82** 031122

Silva C F T d, Posner D and Portugal R 2023 Walking on vertices and edges by continuous-time quantum walk *Quantum Inf. Process.* **22** 16 Paper No. 93

Solenov D 2020 Quantum walks as mathematical foundation for quantum gates *Quantum Inf. Comput.* **20** 230–58

Som A, Sahoo S and Chakrabarti J 2003 Coding DNA sequences: statistical distributions *Math. Biosci.* **183** 49–61

Song L, Zhuge Y, Zuo X, Li M and Wang F 2022 Dna walkers for biosensing development *Advanced Science (Weinheim, Baden-Wurttemberg, Germany)* **9** e2200327

Souza A M C and Andrade R F S 2013a Coin state properties in quantum walks *Sci. Rep.* **3** 1976

Souza A M C and Andrade R F S 2013b Discrete time quantum walk on the Apollonian network *J. Phys. A: Math. Theor.* **46** 13

Souza A M C and Andrade R F S 2019 Fast and slow dynamics for classical and quantum walks on mean-field small world networks *Sci. Rep.* **9** 19143

Staples G S 2008 A new adjacency matrix for finite graphs *Adv. Appl. Clifford Algebras* **18** 979–91

Stefanak M, Jex I and Kiss T 2007 Recurrence and pólya number of quantum walks *Phys. Rev. Lett.* **100** 020501

Strauch F W 2006 Connecting the discrete- and continuous-time quantum walks *Phys. Rev.* A **74** 030301

Sufiani R, Nami S, Golmohammadi M and Jafarizadeh M A 2011 Continuous time quantum walks and quotient graphs *Int. J. Quantum Inf.* **9** 1005–17

Suzuki A 2019 Supersymmetry for chiral symmetric quantum walks *Quantum Inf. Process.* **18** 363

Suzuki A and Tanaka Y 2019 The witten index for 1D supersymmetric quantum walks with anisotropic coins *Quantum Inf. Process.* **18** 377

Tamascelli D and Zanetti L 2014 A quantum-walk-inspired adiabatic algorithm for solving graph isomorphism problems *J. Phys. A: Math. Theor.* **47** 15

Tate T 2022 An eigenfunction expansion formula for one-dimensional two-state quantum walks *Ann. Funct. Anal.* **13** 74

Tonchev H and Danev P 2023 Optimizing the walk coin in the quantum random walk search algorithm *Int. J. Quantum Inf.* **21** 21

Tude L T and de Oliveira M C 2022 Decoherence in the three-state quantum walk *Phys.* A **605** 11

Vakulchyk I, Fistul M V and Flach S 2019 Wave packet spreading with disordered nonlinear discrete-time quantum walks *Phys. Rev. Lett.* **122** 040501

Vakulchyk I, Fistul M V, Qin P and Flach S 2017 Anderson localization in generalized discrete-time quantum walks *Phys. Rev.* B **96** 144204

Valdez F and Melin P 2023 A review on quantum computing and deep learning algorithms and their applications *Soft Comput.* **27** 13217–36

Valero J and Škugor M 2020 Mechanisms, methods of tracking and applications of DNA walkers: a review *Chemphyschem: a European Journal of Chemical Physics and Physical Chemistry* **21** 1971–88

van den Engh G, Sachs R and Trask B J 1992 Estimating genomic distance from DNA sequence location in cell nuclei by a random walk model *Science* **257** 1410–2

Varbanov M, Krovi H and Brun T A 2008 Hitting time for the continuous quantum walk *Phys. Rev.* A **78** 022324

Varsamis G D and Karafyllidis I G 2023 A quantum walks assisted algorithm for peptide and protein folding prediction *Bio Systems* **223** 104822

Varsamis G D, Karafyllidis I G and Sirakoulis G C 2022 Hitting times of quantum and classical random walks in potential spaces *Phys.* A **606** 13

Venegas-Andraca S E 2012 Quantum walks: a comprehensive review *Quantum Inf. Process.* **11** 1015–106

Vitta P B 1969 Modified random-walk problem and the unwinding of DNA *J. Chem. Phys.* **51** 4790–3

Vlachou C, Krawec W O, Mateus P, Paunkovic N and Souto A 2018 Quantum key distribution with quantum walks *Quantum Inf. Process.* **17** 288

Vlachou C, Rodrigues J, Mateus P, Paunković N and Souto A 2015 Quantum walk public-key cryptographic system *Int. J. Quantum Inf.* **13** 10

Vlasov A Y 2018 Effective simulation of state distribution in qubit chains *Quantum Inf. Process.* **17** 20

Wada K 2020 A weak limit theorem for a class of long-range-type quantum walks in 1D *Quantum Inf. Process.* **19** 2

Wagner W 2018 A random walk model for the Schrödinger equation *Math. Comput. Simulation* **143** 138–48

Walczak Z and Bauer J H 2022 Parrondo's paradox in quantum walks with three coins *Phys. Rev.* E **105** 064211

Wald S and Böttcher L 2021 From classical to quantum walks with stochastic resetting on networks *Phys. Rev.* E **103** 012122

Wang C 2022 Higher-dimensional open quantum walk in environment of quantum Bernoulli noises *Stoch. Dynam.* **22** 21

Wang C 2023 Abstract model of continuous-time quantum walk based on Bernoulli functionals and perfect state transfer *Int. J. Quantum Inf.* **21** 19

Wang C, Wang C, Ren S and Tang Y 2018 Open quantum random walk in terms of quantum Bernoulli noise *Quantum Inf. Process.* **17** 14 Paper No. 46

Wang C and Ye X 2016 Quantum walk in terms of quantum Bernoulli noises *Quantum Inf. Process.* **15** 1897–908

Wang F, Zhang P, Wang Y, Liu R, Gao H and Li F 2017 Quantum walk with one variable absorbing boundary *Phys. Lett.* A **381** 65–9

Wang H, Wu J, He H and Tang Y 2016 Localization of two-particle quantum walk on glued-tree and its application in generating Bell states *Quantum Inf. Process.* **15** 3619–35

Wang K, Wu N, Kuklinski P, Xu P, Hu H and Song F 2016 Grover walks on a line with absorbing boundaries *Quantum Inf. Process.* **15** 3573–97

Wang Z-G, Elbaz J and Willner I 2011 Dna machines: bipedal walker and stepper *Nano Lett.* **11** 304–9

Watanabe S, Fukuda A, Segawa E and Sato I 2021 Limit theorem of the max-plus walk *RIMS Kôkyûroku Bessatsu* **B87** 125–33

Willsch M, Willsch D, Michielsen K, Michielsen K, Raedt H D and Raedt H D 2020 Discrete-event simulation of quantum walks *Front. Phys.* **8**

Wing-Bocanegra A and Venegas-Andraca S E 2023 Circuit implementation of discrete-time quantum walks via the shunt decomposition method *Quantum Inf. Process.* **146** 39

Wocjan P and Temme K 2023 Szegedy walk unitaries for quantum maps *Comm. Math. Phys.* **402** 3201–31

Wong T G 2015a Diagrammatic approach to quantum search *Quantum Inf. Process.* **14** 1767–75

Wong T G 2015b Grover search with lackadaisical quantum walks *J. Phys. A: Math. Theor.* **48** 17

Wong T G 2015c Quantum walk search with time-reversal symmetry breaking *J. Phys. A: Math. Theor.* **48** 13

Wong T G 2017 Equivalence of Szegedy's and coined quantum walks *Quantum Inf. Process.* **16** 15

Wong T G and Lockhart J 2021 Equivalent laplacian and adjacency quantum walks on irregular graphs *Phys. Rev.* A **104** 042221

Wong T G and Santos R A M 2017 Exceptional quantum walk search on the cycle *Quantum Inf. Process.* **16** 17

Wong T G, Tarrataca L and Nahimov N 2016 Laplacian versus adjacency matrix in quantum walk search *Quantum Inf. Process.* **15** 4029–48

Wu J, Xu P and Zhu X 2019 Classical random walk with memory versus quantum walk on a one-dimensional infinite chain *Phys. Lett.* A **383** 2389–93

Wu M, Li X, Liu Z and Chen H 2021 Correlation between the continuous-time quantum walk and cliques in graphs and its application *Int. J. Quantum Inf.* **19** 14 Id/No 2 150 009

Xiong D, Thiel F and Barkai E 2017 Using hilbert transform and classical chains to simulate quantum walks *Phys. Rev.* E **96** 022114

Xu M and Tang D 2021 Recent advances in DNA walker machines and their applications coupled with signal amplification strategies: a critical review *Anal. Chim. Acta* **1171** 338523

Xu X-P 2008 Continuous-time quantum walks on one-dimensional regular networks *Phys. Rev. E Stat. Nonlin. Soft Matter Phys.* **77** 061127

Xu X-P 2009 Exact analytical results for quantum walks on star graphs *J. Phys. A: Math. Theor.* **42** 8

Xu X-P 2011 Localization of quantum walks on a deterministic recursive tree *Quantum Inf. Comput.* **11** 253–61

Xu X-P, Li W and Liu F 2008 Coherent transport on apollonian networks and continuous-time quantum walks *Phys. Rev. E Stat. Nonlin. Soft Matter Phys.* **78** 052103

Xu X-P, Zhang X-K, Ide Y and Konno N 2014 Analytical solutions for quantum walks on 1D chain with different shift operators *Ann. Phys.* **344** 194–212

Xu X-Y 2020 Measuring a dynamical topological order parameter in quantum walks *Light Sci. Appl.* **9** 7

Xue P 2012 Implementation of multi-walker quantum walks with cavity grid *J. Comput. Theor. Nanosci.* **10** 1606 1606–12

Xue X-L, Ruan Y and Liu Z-h 2019 Discrete-time quantum walk search on Johnson graphs *Quantum Inf. Process.* **18** 10

Yalçınkaya I and Gedik Z 2015 Qubit state transfer via discrete-time quantum walks *J. Phys. A: Math. Theor.* **48** 12

Yalouz S and Pouthier V 2018 Continuous-time quantum walk on an extended star graph: trapping and superradiance transition *Phys. Rev.* E **97** 11

Yamagami T, Segawa E and Konno N 2021 General condition of quantum teleportation by one-dimensional quantum walks *Quantum Inf. Process.* **20** 1–24

Yamagami T, Segawa E, Mihana T, Röhm A, Horisaki R and Naruse M 2023 Bandit algorithm driven by a classical random walk and a quantum walk *Entropy* **25** 20 Paper No. 843

Yang Y, Bi J, Chen X, Yuan Z, Zhou Y and Shi W 2018 Simple hash function using discrete-time quantum walks *Quantum Inf. Process.* **17** 189

Yang Y-G, Dong J-R, Yang Y-L, Zhou Y-H and Shi W-M 2021 Usefulness of decoherence in quantum-walk-based hash function *Int. J. Theor. Phys.* **60** 1025–37

Yang Y-G, Wang X-X, Li J, Li D, Zhou Y-H and Shi W-M 2021 Decoherence in two-dimensional quantum walks with two- and four-state coins *Mod. Phys. Lett.* A **36** 2150210

Yin R and Barkai E 2023 Restart expedites quantum walk hitting times *Phys. Rev. Lett.* **130** 6 Paper No. 050 802

Yoshie Y 2017 Characterizations of graphs to induce periodic Grover walk *Yokohama Math.* J **63** 9–23

Zaman A, Ahmad R, Bibi S and Khan S 2022 Randomizing quantum walk *Int. J. Theor. Phys.* **61** 10

Zhan H 2019 An infinite family of circulant graphs with perfect state transfer in discrete quantum walks *Quantum Inf. Process.* **18** 369

Zhan H 2021 Quantum walks on embeddings *J. Algebraic Combin* **53** 1187–213

Zhan H 2022 The average search probabilities of discrete-time quantum walks *Quantum Inf. Process.* **21** 25

Zhang Q and Busemeyer J 2021 A quantum walk model for idea propagation in social network and group decision making *Entropy* **23** 19 Paper No. 622

Zhang Y, Song T and Wu Z 2022 An improved algorithm for computing hitting probabilities of quantum walks *Phys.* A **594** 7

Zhao K and Yang W-S 2022 The localization of quantum random walks on Sierpinski gaskets *J. Math. Phys.* **63** 41

Zhao W, Chang Z, Ma C and Shen Z 2023 A pseudorandom number generator based on the chaotic map and quantum random walks *Entropy (Basel, Switzerland)* **25**

Zhou Q and Lu S 2019 One-dimensional quantum walks with two-step memory *Quantum Inf. Process.* **18** 359

www.ingramcontent.com/pod-product-compliance
Ingram Content Group UK Ltd.
Pitfield, Milton Keynes, MK11 3LW, UK
UKHW051851100926
8556IPUK00005B/575

9 780750 352796